ECOLOGY AS POLITICS

ECOLOGY
AS
POLITICS

André Gorz

Translated by Patsy Vigderman and Jonathan Cloud

SOUTH END PRESS, BOSTON

Library of Congress Card Number: 79-64086
ISBN 0-89608-088-9 paper
ISBN 0-89608-089-7 cloth
third printing
Translated from the French by:
    Jonathan Cloud (Part I)
    Patsy Vigderman (Parts II, III, and IV)

Cover design by Ann L. Raszmann/
Boston Community School

Coordinated by Billy Pope

Design, typesetting, and paste up were done by
                South End Press
                116 St. Botolph Street
                Boston, MA 02115

UK Distributor: Pluto Press Limited
Unit 10 Spencer Court/ 7 Chalcot Road
London NW1 8LH

TABLE OF CONTENTS

ECOLOGY AS POLITICS

Introduction

TWO KINDS OF ECOLOGY

Ecology is like universal suffrage or the 40-hour week: at first, the ruling elite and the guardians of social order regard it as subversive, and proclaim that it will lead to the triumph of anarchy and irrationality. Then, when factual evidence and popular pressure can no longer be denied, the establishment suddenly gives way—what was unthinkable yesterday becomes taken for granted today, and fundamentally nothing changes.

Ecological thinking still has many opponents in the board rooms, but it already has enough converts in the ruling elite to ensure its eventual acceptance by the major institutions of modern capitalism.

It is therefore time to end the pretense that ecology is, by itself, sufficient: *the ecological movement is not an end in itself, but a stage in the larger struggle.* It can throw up obstacles to capitalist development and force a number of changes. But when, after exhausting every means of coercion and deceit, capitalism begins to work its way out of the ecological impasse, it will assimilate ecological necessities as technical constraints, and adapt the conditions of exploitation to them.

That is why we must begin by posing the question explicitly: what are we really after? A capitalism adapted to ecological constraints; or a social, economic, and cultural revolution that abolishes the constraints of capitalism and, in so doing, establishes a new relationship between the individual and society and between people and nature? Reform or revolution?

It is inadequate to answer that this question is secondary, and that the main thing is not to botch up the planet to the point where it becomes uninhabitable. For survival is not an end in itself either: is it really worth surviving in a world "transformed into a planetary hospital, planetary school, planetary prison, where it becomes the principal task of spiritual engineers to fabricate people adapted to these conditions"? (Illich)

To be convinced that this is the world which the technocrats are preparing for us, one has only to consider the new "brain-washing" techniques being developed in the U.S. and Germany.[1] Researchers attached to the psychiatric clinic of the University of Hamburg, following the work of American psychiatrists and psychosurgeons, are exploring ways of eliminating the "aggressiveness" which prevents people from accepting the most total forms of frustration—those of the prison system in particular, but also those of the assembly line, of urban crowding, of schooling, red tape, and military discipline.

We should do well, therefore, to define at the outset what we are struggling for, as well as against. And we should do well to try to understand how, concretely, capitalism is likely to be affected and changed by ecological constraints, instead of believing that these will, in and of themselves, bring about its disappearance.

To do this we must first grasp what an ecological constraint means in economic terms. Consider the gigantic chemical plants of the Rhine valley: BASF in Ludwigshafen, AKZO in Rotterdam, or Bayer in Leverkusen. Each of these complexes represents a combination of the following factors:

• natural resources (air, water, minerals) that until now were considered without value and were treated as free goods, because they did not need to be *reproduced* (i.e., replaced);

• means of production (machines, buildings, etc.), i.e., fixed capital, which eventually become obsolete and must consequently be replaced (reproduced), preferably by more efficient and more powerful ones so as to give the firm an advantage over its competitors;

• labor power, which must also be reproduced (the workers must be housed, fed, trained, and kept healthy).

Under capitalism these factors are combined so as to yield the greatest possible amount of profit (which, for any firm interested in its future, means also the maximum control over resources, hence the maximum increase in its investments and presence on the world market). The pursuit of this goal has a profound effect on the way the different factors are combined and the weight given to each.

Corporate management is not, for instance, principally concerned with making work more pleasant, harmonizing production with the balance of nature and the lives of people, or ensuring that its products serve only those ends which communities have chosen for themselves. It is principally concerned with producing the maximum exchange value for the least monetary cost. And to do that it must give greater weight to the smooth running of the machines, which are costly to maintain and replace, than to the physical and psychic health of the workers, who are readily replaceable at low cost. It must give greater weight to lowering the costs of production than to preserving the ecological balances, whose destruction will not burden the firm financially. It must produce what can be sold at the highest prices, regardless of whether cheaper things might be more valuable to the community. Everything bears the imprint of these requirements of capital: the nature of the products, the production technologies, the working conditions, the size and structure of the plants.

But increasingly, and most notably in the Rhine valley, the human crowding and the air and water pollution are reaching the point where industry, in order to grow or even continue operating, is required to filter its fumes and effluents. That is, industry must now reproduce the conditions and resources which were previously considered part of nature and therefore free. This need to reproduce the environment has a chain of economic consequences: it becomes necessary to invest in pollution control equipment, thus increasing the mass of fixed capital; it is then necessary to ensure the amortization (i.e., the reproduction) of the purification installations, but the products of these installations (the restored properties of air and water) cannot themselves be sold for a profit.

In short, there is a simultaneous increase in capital intensity (in the "organic composition" of capital), in the cost of reproducing this fixed capital, and hence in the costs of production, without any corresponding increase in sales. One of two things must therefore occur: either the rate of profit declines or the price of the products increases.

The firm will, of course, seek to raise its prices. But it cannot get off so lightly: all of the other polluting firms (cement plants, steelworks, paper factories, refineries, etc.) will also seek to force the final consumer to pay higher prices for their goods. The incorportation of ecological constraints will in the end have the following results: prices will tend to rise faster than real wages, purchasing power will be reduced, and it will be as if the cost of pollution control had been deducted from the income available to individuals for the purchase of consumer goods. The production of these goods will consequently tend to stagnate or fall off; tendencies towards recession or depression will be accentuated. And this diminution of growth and of production which, in another system, might be considered a positive thing (fewer cars, less noise, more air to breathe, shorter working days, and so on) will instead have entirely negative effects: the polluting goods will become luxury items, inaccessible to the majority but still available to the privileged; inequality will intensify, the poor will become relatively poorer and the rich richer.

Incorporating ecological costs, in short, will have the same social and economic consequences as the oil crisis. And capitalism, far from succumbing to this crisis, will respond to it in the usual fashion: those groups financially advantaged by the crisis will profit from the difficulties of rival groups, will absorb them at a low cost and will extend their control over the economy. The state will reinforce its power over society: its technocrats will calculate "optimal" norms of pollution control and of production, issue regulations, and extend the domain of "programmed" activity and thus the scope of the repressive apparatus. Popular resentment will be diverted with compensatory myths and directed towards readily available scapegoats (racial or ethnic minorities, migrant workers, young people, other countries). The state will base its authority not upon consent but upon coercion; bureaucracies, police forces, armies, and private security forces will fill the vacuum left by the disrepute of party politics and the fossilization of political parties. Already, in France and elsewhere, we see the signs of this decay all around us.

Doubtless, none of this is inevitable. But it is highly probable, if capitalism is compelled to integrate ecological costs *without being challenged* at all levels by alternative social practices and an alternative vision of human civilization. The advocates of growth are right on one point: within the framework of the existing society and consumption patterns—based on disparity, privilege, and the quest for profit—zero or negative growth can only mean stagnation, unemployment, and a widening gap between rich and poor. Within the framework of the existing mode of production, it is impossible to limit or suppress growth while simultaneously distributing goods more equitably.

Indeed, it is the nature of the goods themselves which most often prevents their equitable distribution—how can one equitably distribute supersonic air travel, Mercedes Benzes, penthouse apartments with private swimming pools, or the thousands of new products, scarce by definition, which industry floods the market with each year in order to devalue older models and reproduce inequality and social hierarchy? And how can one "distribute equitably" university degrees, supervisory jobs, managerial roles, or tenured positions?

It is hard to avoid the recognition that the mainspring of growth is this generalized forward flight, stimulated by a deliberately sustained system of inequalities, which Ivan Illich calls "the modernization of poverty." As soon as a majority can aspire to what until then had been the exclusive privilege of the elite, this privilege (the high-school diploma, for example, or the automobile) is thereby devalued, the poverty line is raised by one notch, and new privileges are created from which the majority are excluded. Endlessly re-creating scarcity in order to re-create inequality and hierarchy, capitalist society gives rise to more unfilfilled needs than it satisfies: "the growth-rate of frustration largely exceeds that of production." (Illich)

As long as we remain within the framework of a civilization based on inequality, growth will necessarily appear to the mass of the people as the promise—albeit entirely illusory—that they will one day cease being "under privileged," and the limitation of growth as the threat of permanent mediocrity. It is not so much growth that must be attacked as the illusions which it sustains, the dynamic of ever-growing and ever-frustrated needs on which it is based, and the competition which it institutionalizes by inciting each individual to seek to rise "above" all others. The motto of our society could be: *That which is good for everyone is without value; to be respectable you must have something "better" than the next person.*

Now it is the very opposite which must be affirmed in order to break with the ideology of growth: *The only things worthy of each are those which are good for all; the only things worthy of being produced are those which neither privilege nor diminish anyone; it is possible to be happier with less affluence, for in a society without privilege no one will be poor.*

Imagine a society based on these criteria: the production of practically indestructible materials, of apparel lasting for years, of simple machines which are easy to repair and capable of functioning for a century or more. These are scientifically and technically accessible to us, along with a vast extension of community services and facilities (public transportation, laundromats, etc.), thus eliminating much of the need for fragile, expensive, and energy-wasting private machines.

Imagine collective dwellings—not, as they now are, blighted by the neglect of public space and the privatization of value, but as they might be if individual energies were to be released for the public good and vice versa. There could be two or three recreation rooms, playrooms for children, fully-equipped workshops and libraries, and accessible laundry areas. Would we still really require all of our individual equipment which lies idle much of the time? Would we still be eager to join the traffic jams on the freeways if there were comfortable, collective transport to recreation areas, bicycles and motorbikes readily available when required, an extensive network of mass transit for urban and suburban areas?

Imagine beyond this that the major industries, centrally planned, produced only that which was required to meet the basic needs of the population: four or five styles of durable shoes and clothing, three or four models of sturdy and adaptable vehicles, plus everything needed to provide the collective services and facilities. Impossible in a market economy? No doubt. Entailing massive unemployment? Not necessarily. We could have a 20-hour work week, providing we change the system. Uniformity, monotony, boredom? On the contrary, for imagine the following:

Each neighborhood, each town, would have public workshops equipped with a complete range of tools, machines, and raw materials, where the citizens *produce for themselves, outside the market economy*, the non-essentials according to their tastes and desires. As they would not work more than twenty hours a week (and possibly less) to produce the necessities of life, the adults would have time to learn what the children would be learning in primary school: not only reading and writing but also handicrafts of all kinds, sewing, leather-working, cabinet-making, masonry, metal-working, mechanics, pottery, agriculture—in short, all of the skills which are now commercially torn from us and replaced with buying and selling.

Are such proposals utopian? Why couldn't they become a political program? For such a "utopia" corresponds to the most advanced, not the most primitive, form of socialism—to a society without bureaucracy, where the market withers away, where there is enough for everyone, where people are collectively and individually free to shape their lives, where people produce according to their fantasies, not only according to their needs; in short, a society where "the free development of each is the condition for the free development of all." (Marx, *The Communist Manifesto*, 1848)

Le Sauvage, April 1974

NOTES

1. Documented in *Les Temps Modernes*, March 1974.

Chapter 1

ECOLOGY AND FREEDOM

1. Ecological Realism

Growth-oriented capitalism is dead. Growth-oriented social-ism, which closely resembles it, reflects the distorted image of our past, not of our future. Marxism, although irreplaceable as an instrument of analysis, has lost its prophetic value.

The development of the forces of production, which was supposed to enable the working class to cast off its chains and establish universal freedom, has instead dispossessed the work-ers of the last shreds of their sovereignty, deepened the division between manual and intellectual labor, and destroyed the mate-rial and existential bases of the producers' power.

Economic growth, which was supposed to ensure the afflu-ence and well-being of everyone, has created needs more quickly than it could satisfy them, and has led to a series of dead ends which are not solely economic in character: capitalist growth is in crisis not only because it is capitalist but also because it is encountering physical limits.

It is possible to imagine palliatives for one or another of the problems which have given rise to the present crisis. But its distinctive character is that it will inevitably be aggravated by each of the successive and partial solutions by which it seeks to overcome them.

While it has all the characteristics of a classical crisis of over-production, the current crisis also possesses a number of new dimensions which Marxists, with rare exceptions, have not foreseen, and for which what has until now been understood as "socialism" does not contain adequate answers. It is a crisis in the relation between the individual and the economic sphere as such; a crisis in the character of work; a crisis in our relations with nature, with our bodies, with our sexuality, with society, with future generations, with history; a crisis of urban life, of habitat, of medical practice, of education, of science.

We know that our present mode of life is without future; that the children we will bring into the world will use neither oil nor a number of now-familiar metals during their adult lives; that if current nuclear programs are implemented, uranium reserves will be exhausted by then.

We know that *our* world is ending; that if we go on as before, the oceans and the rivers will be sterile, the soil infertile, the air unbreathable in the cities, and life a privilege reserved for the selected specimens of a new race of humans, adapted by chemical conditioning and genetic programming to survive in a new ecological niche, carved out and sustained by biological engineering.

We know that for a hundred and fifty years industrial society has developed through the accelerated looting of reserves whose creation required tens of millions of years; and that until very recently all economists, whether classical or Marxist, have rejected as irrelevant or "reactionary" all questions concerning the longer-term future—that of the planet, that of the biosphere, that of civilizations. "In the long run we shall all be dead," said Keynes, wryly asserting that the temporal horizon of the economist should not exceed the next ten or twenty years. "Science," we were assured, would find new paths; engineering would discover new processes undreamt of today.

But science and technology have ended up making this

central discovery: all productive activity depends on borrowing from the finite resources of the planet and on organizing a set of exchanges within a fragile system of multiple equilibriums.

The point is not to deify nature or to "go back" to it, but to take account of a simple fact: human activity finds in the natural world its external limits. Disregarding these limits sets off a backlash whose effects we are already experiencing in specific, though still widely misunderstood, ways: new diseases and new forms of dis-ease, maladjusted children (but maladjusted to what?), decreasing life expectancy, decreasing physical yields and economic pay-offs, and a decreasing quality of life despite increasing levels of material consumption.

The response of economists up to now has essentially consisted of dismissing as "utopian" or "irresponsible" those who have focused attention on these symptoms of a crisis in our fundamental relation to the natural world, a relation in which all economic activity is grounded. The boldest concept which modern political economy dared envisage was that of "zero growth" in physical consumption. Only one economist, Nicholas Georgesco-Roegen, has had the common sense to point out that, even at zero growth, the continued consumption of scarce resources will inevitably result in exhausting them completely. The point is not to refrain from consuming more and more, but to consume less and less—there is no other way of conserving the available reserves for future generations.

This is what ecological realism is about.

The standard objection is that any effort to arrest or reverse the process of growth will perpetuate or even worsen existing inequalities, and result in a deterioration in the material conditions of those who are already poor. But the idea that growth reduces inequality is a faulty one—statistics show that, on the contrary, the reverse is true.[1] It may be objected that these statistics apply only to capitalist countries and that socialism would produce greater social justice; but why then should it be necessary to produce more things? Would it not be more rational to improve the conditions and the quality of life by making more efficient use of available resources, by producing different things differently, by eliminating waste, and by refusing to produce socially those goods which are so expensive that they can never be available to all, or which are so cumbersome or polluting that their costs outweigh their benefits as soon as they become accessible to the majority?[2]

Radicals who refuse to examine the question of equality without growth merely demonstrate that "socialism," for them, is nothing but the continuation of capitalism by other means— an extension of middle class values, lifestyles, and social patterns (which the more enlightened members of that class, under pressure from their daughters and sons, are already beginning to reject).

Today a lack of realism no longer consists in advocating greater well-being through the inversion of growth and the subversion of the prevailing way of life. Lack of realism consists in imagining that economic growth can still bring about increased human welfare, and indeed that it is still physically possible.

2. Political Economy and Ecology: Marx and Illich

Political economy, as a specific discipline, applies neither to the family nor to those communities small enough to settle by common agreement the cooperation of their members and their exchange (or pooling) of goods and mutual services. Political economy begins only where free cooperation and reciprocity cease. It begins only with *social production*, i.e., production founded upon a social division of labor and regulated by mechanisms external to the will and consciousness of individuals—by market processes or by central planning (or by both).

"Economic man," i.e., the abstract individual who underpins economic reasoning, has the unique characteristic of not consuming what he or she produces and not producing what he or she consumes. Consequently he or she is never troubled by questions of quality, usefulness, charm, beauty, happiness, freedom, or morality, but is affected only by exchange values, flows, and quantitative aggregates and balances.

Economists do not concern themselves with what individuals think, feel, and desire, but only with the material processes which, independently of their own will, human activities give rise to in a (social) context of limited resources.

It is impossible to derive an ethic from economic reasoning. Marx was one of the first to understand this. The choice he discerned was, very schematically, as follows:

• either individuals manage to unite and, in order to subordinate the economic process to their collective will, replace the social division of labor with the voluntary cooperation of associated producers;

• or else they remain dispersed and divided, in which case the economic process will prevail over people's aims and goals, and sooner or later a strong central state will, in the pursuit of its own rationality, impose by force the cooperation which the people were unable to achieve for themselves. The choice is simple: "socialism or barbarism."

The ecologist stands in the same relation to economic activity as the economist to the convivial cooperation which rules family or community activities. Ecology, as a specific discipline, does not apply to those communities or peoples whose ways of producing have no lasting or irremediable effects on the environment—natural resources appear inexhaustible, the impact of human activity negligible. In the ideal case, the stewardship of nature is, like the art of healthy living, based on the unwritten rules of generally accepted wisdom.

Ecology does not appear as a separate discipline until economic activity destroys or permanently disturbs the environment and, in so doing, compromises the pursuit of economic activity itself, or significantly changes its conditions. Ecology is concerned with the external limits which economic activity must respect so as to avoid producing effects contrary to its aims or incompatible with its continuation.

In the same way that economics is concerned with the external constraints that *individual* activities give rise to when they generate unwanted *collective* results, ecology is concerned with the external constraints which economic activity gives rise to when it produces environmental alterations which upset the calculation of costs and benefits.

In the same way that economics belongs to a realm beyond reciprocity and voluntary cooperation, ecology belongs to a realm beyond that of economic activity and calculation, *but without including it*—it is not the case that ecology is a superior rationality which subsumes that of economics. Ecology has a different rationality: it makes us aware that the efficiency of economic activity is a limited one and that it rests upon extra-economic conditions. It enables us to discover, in particular, that the economic effort to overcome *relative* scarcities engenders, beyond a certain threshold, *absolute and insurmountable* scarcities. The returns become negative: production destroys more than it produces. This inversion occurs when economic activity infringes upon the equilibrium of primary ecological cycles and/or destroys resources which it is incapable of regenerating or reconstituting.

To this type of situation, the economic system has in the past invariably responded by additional productive efforts; it tries to compensate with increased production for the scarcities engendered by increased production. It does not recognize that this response necessarily exacerbates these scarcities: that, beyond a certain threshold, measures favoring the circulation of automobiles increase congestion; that the increased consumption of medicine increases morbidity while displacing its causes; that the increased consumption of energy creates forms of pollution which, as long as they remain uncontrolled at their source, can only be fought in ways which involve a new increase in energy consumption, itself polluting, and so on.

To understand and overcome these "counterproductivities," one has to break with economic rationality.[3] This is what ecology does: it reveals to us that an appropriate response to the scarcities and disease, to the bottlenecks and dead-ends of industrial civilization, must be sought not in growth but in the limitation or reduction of material production. It demonstrates that it can be more effective and "productive" to conserve natural resources than to exploit them, to sustain natural cycles rather than interfere with them.

It is nevertheless impossible to derive an ethic from ecology. Ivan Illich is one of the first to have understood this. The alternatives which he sees before us can be stated schematically as follows:

● either we agree to impose limits on technology and industrial production so as to conserve natural resources, preserve the ecological balances necessary to life, and favor the development and autonomy of communities and individuals (this is the convivial option);

• or else the limits necessary to the preservation of life will be centrally determined and planned by ecological engineers, and the programmed production of an "optimal" environment will be entrusted to centralized institutions and hard technologies (this is the technofascist option, the path along which we are already halfway engaged).[4] The choice is simple: "conviviality or technofascism."

Ecology, as a purely scientific discipline, does not necessarily imply the rejection of authoritarian, technofascist solutions. The rejection of technofascism does not arise from a scientific understanding of the balances of nature, but from a political and cultural choice. Environmentalists *use* ecology as the lever to push forward a radical critique of our civilization and our society. But ecological arguments can also be used to justify the application of biological engineering to human systems.

3. Ecology and the Inversion of Tools[5]

The preference for natural, self-regulating systems over systems relying on experts and institutions need not imply a quasi-religious exaltation of nature. It is not impossible for artificial systems to be, in certain respects, more efficient than natural ones. The preference for the latter should be defended as a *rational choice*, in both political and ethical terms—a preference for decentralized self-regulation over centralized other-regulation. The field of "health policy" provides us with a particularly striking example, which can serve as a paradigm.

Natural selection is the perfect case of decentralized self-regulation. It can be circumvented by the increasingly sophisticated interventions of the medical-care apparatus, which can save the lives of babies who would otherwise die in their first days or months. These individuals, however, will in turn tend to have offspring of whom a growing proportion will display hereditary defects or diseases. The resulting deterioration of the genetic stock is already leading some geneticists to advocate a state-enforced policy of eugenics—that is, a regulation of the freedom to mate and procreate.

The abolition of natural self-regulation thus leads to the necessity for administrative regulation. Natural selection is in the end to be replaced by social selection.

The latter can, in certain respects, be regarded as more efficient than the former: eugenics would prevent the conception of deformed or non-viable individuals, whereas natural selection eliminates them only after conception or, often, only after birth. But there is another difference: natural selection occurs spontaneously, without any planned intervention. Eugenics, on the other hand, assumes a technobureaucracy capable of enforcing the administrative norms which it lays down. Natural self-regulation can only be replaced by regulating *authority*.

This example, in no way fanciful, is intended to illustrate the ecological principle that *it is better to leave nature to work itself out than to seek to correct it at the cost of a growing submission of individuals to institutions, to the domination of others.* For the ecologist's objection to system engineering is not that it violates nature (which is not sacred), but that it substitutes new forms of domination for existing natural processes.

Politically, the implication is obvious: the ecological perspective is incompatible with the rationality of capitalism.[6] It is also wholly incompatible with the authoritarian socialism which (whether it relies on central economic planning or not) is the only kind which exists in the world today on a governmental level. The ecologist's position is not, by contrast, incompatible with a libertarian or democratic socialism: but it should not be confused with it. The ecologist's concern is working at another and more fundamental level: that of the material prerequisites of the economic system. In particular, it is concerned with the character of prevailing technologies, for the techniques on which the economic system is based are not neutral. In fact, they reflect and determine the relations of the producers to their products, of the workers to their work, of the individual to the group and the society, of people to the environment. Technology is the matrix in which the distribution of power, the social relations of production, and the hierarchical division of labor are embedded.[7]

Societal choices are continually being imposed upon us under the guise of technical choices.[8] These technical choices are rarely the only ones possible, nor are they necessarily the most efficient ones. For capitalism develops only those technologies which correspond to its logic and which are compatible with its continued domination. It eliminates those technologies which do not strengthen prevailing social relations, even where they are more rational with respect to stated objectives.[9] Capitalist relations of production and exchange are already inscribed in the technologies which capitalism bequeaths to us.

The struggle for different technologies is essential to the struggle for a different society. The institutions and structures of the state are to a large extent determined by the nature and weight of its technologies. Nuclear energy, for example—whether "capitalist" or "socialist"—presupposes and imposes a centralized, hierarchical, police-dominated society.

The inversion of tools is a fundamental condition of the transformation of society. The development of voluntary cooperation, the self-determination and freedom of communities and individuals, requires the development of technologies and methods of production which:

• can be used and controlled at the level of the neighborhood or community;

• are capable of generating increased economic autonomy for local and regional collectivities;

• are not harmful to the environment; and

• are compatible with the exercise of joint control by producers and consumers over products and production processes.

Of course, it can be objected that it is impossible to change the tools without transforming society as a whole, and that this cannot be accomplished without gaining control over the state. This objection is valid providing it is not taken to mean that societal change and the acquisition of state power must *precede* technological change. For without changing the technology, the transformation of society will remain formal and illusory. The theoretical and practical definition of alternative technologies, and the struggle of communities and individuals to win, collectively and individually, control over their own destinies, must be the permanent focus of political action. If they are not, the seizure of state power by people calling themselves socialists will not change fundamentally either the system of domination or the relations of men and women to each other and to nature. Socialism is not immune to technofascism. It will, on the contrary, fall prey to it whenever and wherever it sets out to enhance and multiply the powers of the state without developing simultaneously the autonomy of civil society.

This is why the ecological struggle is, in its present form, an indispensable dimension of the struggle against capitalism. It cannot be subordinated to the political objectives of socialism. Only where the left is committed to a fully decentralized and democratic socialism can it give political expression to ecological demands. The organized left, in France as in other countries, has not yet reached this stage; it has not incorporated ecological principles in either its practice or its program. It is for this reason that the ecological movement must continue to assert its specificity and its autonomy.

Ecological concerns are fundamental; they cannot be compromised or postponed. Socialism is no better than capitalism if it makes use of the same tools. The total domination of nature inevitably entails a domination of people by the techniques of domination. If there were no other options, it would be preferable to have a non-nuclear capitalism than to have a nuclear socialism, for the former would weigh less heavily upon future generations.

4. Ecology and the Crisis of Capitalism

All production is also destruction. This fact can be overlooked so long as production does not irreversibly deplete natural resources: resources may then appear inexhaustible. They regenerate themselves naturally—the grass grows back, along with the weeds. The effects of destruction appear wholly productive. More precisely: this destruction is the very condition of production. It has to be repeated again and again.

This process is unavoidable. The earth is not naturally hospitable to humankind. Nature is not a garden planted for our benefit. Human life on earth is precarious and, in order to expand, it must displace some of the natural equilibriums of the ecosystem. Agriculture is the first organized expression of this: it alters not only the balance between plant species but also between plant and animal ones. In particular, it entails a struggle against pests and diseases, a struggle which can be carried out by biological as well as chemical means—that is, by favoring certain species, considered "desirable," so that they will control others, considered "undesirable." In this way agriculture reshapes the surface of the earth.

Nature is not untouchable. The "promethean" project of "mastering" or "domesticating" nature is not necessarily incompatible with a concern for the environment. All culture (in the double sense of this word) encroaches upon nature and modifies the biosphere. The fundamental issue raised by ecology is simply that of knowing:

• whether the exchanges, which human activity imposes upon or extorts from nature, preserve or carefully manage the stock of nonrenewable resources; and

• whether the destructive effects of production do not exceed the productive ones by depleting renewable resources more quickly than they can regenerate themselves.

On both counts, there is little doubt that ecological factors play a determining and aggravating role in the current economic crisis. This does not mean that these factors should be regarded as the primary causes of the crisis: we are dealing, rather, with a crisis of capitalist overaccumulation, intensified by an ecological crisis (and, as we shall see, by a social one).

To make this clearer, I shall deal separately with the different levels of this crisis:

a. *The crisis of overaccumulation.* In its advanced stages, the development of capitalism rests principally on the replacement of workers by machines, of living labor by dead labor. The machine, in effect, is work that has been accumulated and embodied in an inert, inanimate form capable of being expended in the absence of the worker. But machines are costly to produce; the investment of capital which they represent must be profitable, which means: the investor expects a return greater than the cost of the installation. Insofar as it serves to produce this surplus, *through the mediation of the workers who operate it*, the machine is capital. The logic of capital is the pursuit of constant growth.

Grow or perish, that is the law of capital. Except in periods of prolonged stagnation, when the firms in a given sector reach an agreement to share the market and charge the same prices (which is usually called a cartel), the various enterprises compete with one another. They do this in the following manner: each tries to make its machines pay off as quickly as possible so as to be able to install even more efficient ones—machines which can produce the same volume of production with a smaller number of workers. This is called "increasing productivity."

Thus, as advanced capitalism develops, more and more sophisticated and costly machines are operated by fewer and fewer workers, who are less and less skilled. In the costs of production the share of direct wages decreases, while the share of capital increases (which is to say: the amount of profit which must be made in order to pay off and renew the machines increases). In Marxist terminology, the "organic composition of capital" increases. Another way of describing this is to say that industry becomes more and more capital-intensive: it uses an increasing amount of capital to produce the same volume of commodities. It must therefore produce a larger mass of profits to replace and renew the machines, while at the same time compensating the investment capital (in large part loaned by banks) at a rate of interest satisfactory to the lenders.

Marx demonstrated that, sooner or later, the average rate of profit must decline: the more capital is used to produce the same volume of commodities, the more the profit which can be derived from this production diminishes in relation to the mass of capital employed. In other words, this mass cannot keep increasing without eventually reaching a limit.

But from the moment that the rate of profit begins to decline, the whole system is jammed: the machines cannot be made to turn out goods which yield the usual profit, nor, consequently, can they be replaced at the same rate as before; hence the production machinery (amongst other things) begins to fall off and the decline in production progressively spreads. In Marxist terms, there is "overaccumulation": the share of capital in production has become so great (its organic composition so heightened) that it cannot reproduce itself at a normal rate. The "productivity" of capital declines. The value of the fixed capital, which cannot be made to yield a sufficient profit, declines to zero. This capital will, in fact, be destroyed: machines are discarded, factories closed down, workers laid off. The system is in crisis.

In order to avoid this crisis, the managers of capitalism are constantly forced to work against the tendency of the rate of profit to fall off. There are basically two means available to them:

• increasing the quantity of goods sold;

• increasing not the quantity but the price (the exchange value) of the goods sold, e.g., by making them more elaborate and sophisticated.

These two approaches are obviously not mutually exclusive. In particular, it is possible to increase sales by making products less durable, thus forcing people to change them more often; at the same time, these products can be made more complicated and expensive.

This is the nature of consumption in affluent societies; it ensures the growth of capital without increasing either the level of general satisfaction or the number of genuinely useful goods ("use values") which people have at any given point in time. On the contrary: it requires an increasing quantity of products to provide the same level of need-satisfaction. Increasing amounts of energy, of labor, of raw materials, and of capital are "consumed" without people being significantly better off. Production becomes more and more destructive and wasteful; the destruction or obsolescence of products is built into them—their rapid deterioration is programmed.

Thus we have seen tin cans replaced by aluminum ones, which require fifteen times as much energy to produce; rail transport replaced by road transport, which consumes six to seven times as much energy, and uses vehicles which must be replaced more often; the disappearance of objects assembled with bolts and screws in favor of welded or molded ones, which are thus impossible to repair; the reduction of the durability of stoves and refrigerators to around six or seven years; the replacement of natural fibers and leather with synthetic materials which wear out faster; the extension of disposable packaging, which wastes as much energy as non-returnable glass; the introduction of throwaway tissues and dishes; the widespread construction of skyscrapers of glass and aluminum, which consume as much energy for cooling and ventilation in the summer as for heating in the winter; and so on.

Predictably, this type of growth turned out to be a forward flight, not a lasting solution. Advanced capitalism sought to avoid falling rates of profit and the saturation of markets by an accelerated circulation of capital and the planned obsolescence of consumer products.[10] We shall see that it thereby created effects contrary to its original objectives (which economists call "side-effects" or "disutilities") while at the same time generating new relative scarcities, new dissatisfactions, and new forms of poverty.

This forward flight, which was in any event bound to culminate in economic crisis, came to a stop with the so-called oil crisis. The latter did not cause the economic recession; it merely revealed and aggravated the recessionary tendencies which had been brewing for several years. Above all, the oil crisis revealed the fact that capitalist development had created absolute scarcities: in trying to overcome the economic obstacles to growth, capitalist development had given rise to physical obstacles.

b. *The crisis of reproduction.* Under capitalism, absolute scarcity is normally reflected in soaring prices before it appears as physical shortage. According to the dogma of liberal (or neoclassical) economics, the rising price of a scarce good results in the increased production of that good, for this production is becoming more profitable. This line of reasoning assumes however that the scarce good is always *producible.* But the scarcities which have appeared or been aggravated since the middle of the 1960s are principally those of *non-producible* goods. Increased human activity is not capable of making these goods available in greater quantities: they are scarce because they are to be found only in limited quantities in nature.

This applies to the availability of land in heavily indus-trialized areas; to air, water, and the natural fertility of the soil; to forests, fisheries, and an increasing number of raw materials. The explosive rise in prices served only to aggravate the econo-mic crisis or rather to hasten its arrival, for it contributed to the falling rate of profit in two ways:

• When air, water, and urban land become scarce, it is impossible to produce greater quantities of them no matter what price is assigned to them. They can only be shared or redistri-buted in a different way. As far as land is concerned, this means building highrises or underground, or paying higher and higher prices for agricultural land on which to build factories, cities, and roads. In the case of air and water, it means that the available supply must be recycled. This has become necessary not only in Japan but also in the Rhine valley: the German chemical industry has had to forego expansion because the investments required for atmospheric recycling would have been too great.[11]

The need for such recycling has a precise economic signifi-cance: it means that from now on it has become necessary to *reproduce* that which was previously abundant and free. Air and water, in particular, have become means of production like any others: industries must now assign a portion of their invest-ments to antipollution equipment in order to restore to the air and water some of their original properties. The consequence of this requirement is a further increase in the organic composition of capital (i.e., in the share of capital per amount of commodi-ties produced). But there is no corresponding increase in the amount of merchandise produced; the air and water recycled or depolluted by the chemical industry cannot be resold. The fall-ing rate of profit is thus aggravated; the productivity of capital encounters *physical limits*. And those limits created by pollution are not the only ones.

• The exhaustion of the most accessible mineral deposits, i.e., those which cost the least to exploit, constitutes a second physical limitation to the ability of industrial capital to return a profit. In effect, new deposits of raw materials cannot be dis-covered and exploited except at the cost of higher investments than in the past. The financing of these investments implies a higher price for primary products; the higher price of such products in turn bears on the profits of the manufacturing industries at a time when these are tending to decline for the reasons already outlined.

Moreover, mineral prospecting and extraction will in the future require even heavier investment than at present. In view of the rapidly rising prices of the raw material recovered at these higher costs, processing industries must begin developing new technologies which make more efficient use of primary products, including energy. This also requires further investments.

This helps to explain the original and seemingly paradoxical characteristics of the present crisis: despite existing over-capacities, despite the declining rate of profit, and despite the recession, investment remains at an unusually high level and prices continue to rise. Traditional economic reasoning is incapable of accounting for this paradox, which only becomes intelligible when looked at in terms of the underlying physical realities.

Capital, under these conditions, encounters unavoidable difficulties in financing further investments—it becomes incapable of ensuring its own reproduction. The replacement of industrial capital (which is to say, *grosso modo,* of the physical apparatus of production) can no longer be accomplished by the transfer of a surplus levied upon consumption—the reproduction of the system simply costs more than it yields. In other words, *industry consumes more for its own needs:* it delivers fewer products to the final consumer than it used to. Its efficiency has diminished; its physical costs have increased. This is where we are today.

The chain of events which led up to this situation can be broken down into two principal phases:

• During the first phase, production becomes increasingly wasteful, i.e., destructive, in order to avoid a crisis of over-accumulation. It speeds up the destruction of the non-renewable resources on which it depends; and it overconsumes resources which are in principle renewable (air, water, forests, soil, etc.) at a pace which rapidly renders them scarce as well.

• During the second phase, confronted with the depletion of pillaged resources, industry makes frantic efforts to overcome the scarcities engendered by increased production by further increasing production. But the products of this additional production are not added to final consumption; they are consumed by industry itself.

From the point of view of the final consumer, it is as if industry has to produce more—and hence to consume more, in the form of wealth and resources—in order to maintain the same level of consumption for the population. The balance between production and consumption is shifted at the expense of the latter. The overall efficiency of the system goes down. The altering of property relations (i.e., by nationalization) is incapable of remedying this decline in efficiency. It can at most— and during a limited period—facilitate the transfer of resources from consumption to investment. *But nationalization cannot initiate a new phase of sustained growth in material consumption.* For the obstacles to growth have become substantive ones.

In summary, we are dealing with a classical crisis of over-accumulation, aggravated by a crisis of reproduction which is due, in the final analysis, to the increasing scarcity of natural resources. The solution to this crisis cannot be found in the recovery of economic growth, but only in an inversion of the logic of capitalism itself. This logic tends intrinsically towards maximization: creating the greatest possible number of needs and seeking to satisfy them with the largest possible amount of marketable goods and services in order to derive the greatest possible profit from the greatest possible flow of energy and resources. But the link between "more" and "better" has now been broken. "Better" may now mean "less": creating as few needs as possible, satisfying them with the smallest possible expenditure of materials, energy, and work, and imposing the least possible burden on the environment.

This can be done without impoverishment or social injustice, without reducing the quality of life, providing we are prepared to attack the source of poverty. This source is not the lack of production as such but the nature of the goods produced, the pattern of consumption which capitalism promotes, and the inequality which drives it. I shall try to show this in greater detail in the following two sections.

5. The Poverty of Affluence

A richer life is not only compatible with the production of fewer goods, it demands it. Nothing—other than the logic of capitalism—prevents us from manufacturing and making available to everyone adequate accomodation, clothing, household equipment, and forms of transportation which are energy-conserving, simple to repair, and longlasting, while simultaneously increasing the amount of free time and the amount of truly useful products available to the population.

The connection between "living better" and "producing less" seems to be already well understood by a large segment of the population. In France, according to one recent survey:

• 53% of the population would accept a reduction in growth and material consumption, providing it was coupled with changes in lifestyles;

• 68% would prefer more classical and longer-lasting clothes to those which must be discarded after a single season;

• 75% consider throwaway packaging and non-returnable containers needlessly wasteful;

• 78% would welcome one night a week without television as an opportunity to spend time with each other and have face-to-face conversations.[12]

In industrially advanced societies, people do not stay poor for lack of a large enough supply of consumer goods, but because of the nature of the goods produced and the way of producing them. To eliminate poverty we no longer need larger quantities of goods but only different goods, to be produced in a different way.

The *persistence* of poverty in advanced industrial societies cannot be ascribed to the same factors as the *existence* of poverty in the so-called underdeveloped countries. Whereas the latter can, in the final analysis, be attributed to physical shortages, which can be overcome by the development (under specified conditions) of the forces of production, the persistence of poverty in rich countries must be attributed to a social system which produces poverty at the same time it produces increasing wealth. Poverty is created and maintained, that is to say *produced and reproduced*, at the very pace at which the level of aggregate consumption rises.

Before explaining the mechanisms underlying this reproduction, it is important to recognize that the scarcity of natural resources is not experienced in the same way when these resources are equitably distributed as when their distribution is inequitable. Marshall Sahlins has convincingly demonstrated that poverty and inequity are mutually exclusive: physical scarcity, as in the so-called primitive societies, may create frugality or even utter destitution, but it cannot cause "poverty" as long as those resources which do exist are equally accessible and distributed to everyone.[13] Poverty entails, by definition, *the privation of wealth available to others*: the rich. Just as there are no poor when there are no rich, so there can be no rich when there are no poor: when everyone is "rich" no one is, and the same is true for poverty. *As opposed to destitution, which refers to a shortage of the necessities of life, poverty is essentially a relative condition.*

Following these definitions, we can distinguish three major causes of poverty in industrialized societies:

a. *Detrimental appropriation (accaparement).* This is the most obvious cause of poverty: the rich monopolize resources which would otherwise be available in sufficient quantities for all. A typical instance of this is the amassing of land and water rights where these are in principle sufficient to meet the needs of everyone—the equitable distribution of such resources is openly denied. The monopolization of these resources by the few cannot be accounted for by the fact of scarcity—which, on the contrary, follows from it—but only by the domination of one class or caste over another.[14]

b. *Exclusive access.* We speak of exclusive access when the dominant minority bars the rest of the people from access to those naturally-occurring resources which, either because of their scarcity or because of their intrinsic character, cannot be equally distributed or made available to everyone at the same time. A typical example is the establishment of rights of access to natural areas whose attractiveness might be destroyed if "invaded" by large numbers of people; or exacting a price for natural amenities such as clean air, natural light, or silence, which cannot be preserved in a given location without restricting access to it.

The establishment of exclusive rights is most often achieved by the industrialization of access:[15] to get access to a beach, one has to rent a hotel room, buy a meal in a beach restaurant, or purchase a villa; to enjoy sunlight or quiet, it may be necessary to rent or purchase a dwelling which is more expensive because of the limited availability of these resources, although they are in themselves free.[16]

In these instances, exclusive access does not itself create the scarcity—scarcity is real, and there may be no remedy for it. Exclusive access must not, therefore, be regarded as an ultimate obstacle to equitable distribution; it really preserves something which if equally distributed would disappear, and for which, therefore, such distribution is not possible. But this preservation is accomplished, in most societies, for the exclusive benefit of a minority for whom this exclusive access also constitutes a symbol of wealth and power.

The example of the availability of light and quiet also demonstrates the possibility of developing new inequalities—and thereby new divisions between the rich and the poor—by creating *artificial scarcities* in otherwise abundant resources. This creation of artificial scarcities is one of the principal mechanisms by which poverty is reproduced. By destroying, without apparent necessity or advantage, previously abundant resources, and then instituting rights of access to or commercializing those which remain, capitalism creates new forms of privilege and poverty, and prevents the elimination of poverty conditions.

c. *Distinctive consumption.* We use this term to refer to the consumption of goods and services of doubtful use value but which, because of their limited availability or high price, confer status or prestige upon those who have access to them. Distinctive consumption may entail detrimental appropriation, but need not invariably do so. Travelling by supersonic aircraft, for example, involves a detrimental and wasteful expenditure of resources. The Concorde represents the detrimental use of a huge quantity of labor which, in principle, could have been devoted to purposes beneficial to the society as a whole; moreover, each flight involves the detrimental appropriation of large quantities of fuel, thus contributing to the further depletion of the world's oil reserves.

The Concorde is at the same time a source of poverty independent of the detrimental appropriation of social resources which it implies: it conspicuously demonstrates the inequality of desires and of the power to fulfill them. The desire to fly at twice the speed of sound in order to save four hours between Paris and New York is above all the desire for something exceptional, which designates as exceptionally important and powerful those who obtain it. People who utilize this means of transport do not choose it simply for the pleasure or benefits which it provides (subsonic travel is in fact more comfortable) but to assert their distinctive right to a scarce good, reserved by definition to the privileged and the powerful.

Distinctive consumption is the second major mechanism involved in the reproduction of poverty. Once a product enters into common usage, it is time to launch a new product. The product, which is initially scarce and expensive because of its very novelty, enables the rich—independently of all superiority of the new product over the old one—to distinguish themselves *as* rich and to reestablish the poverty of the poor. This is again what Ivan Illich calls "modernization of poverty."[17]

The elimination of poverty will thus never be accomplished by increasing production. What is required is a reorientation of production according to the following criteria:

- socially produced goods must be available to everyone;
- their production must not entail the destruction of naturally abundant resources;
- they must be designed in such a way that, by becoming available to all, they do not cause pollution or bottlenecks which destroy their use value.

But that's not all. The reorientation of production to conform to these criteria also presupposes a "cultural revolution": poverty will only disappear if the inequalities of power and rights, which are its principal source, are also eradicated. Indeed, differences in consumption are often no more than the *means* through which the hierarchical nature of society is expressed. In extreme cases, the one and only purpose of distinctive consumption is to constitute others as poor, not to acquire anything which is intrinsically desirable. This is the case, for example, in the consumption of precious stones or high fashion articles. These conspicuous goods do not even procure pleasure, power, or comfort: they simply demonstrate the power of acquiring things which are beyond the reach of others. The only function of these things is to make social inequality tangible.

Consequently, equality in consumption can only be the result of, and not a means to, the achievement of social equality. The latter depends upon the abolition of hierarchical order. If a hierarchy of powers and functions persists, it will soon reestablish both material and symbolic inequalities (as has occurred in authoritarian socialist societies). If it is abolished, material inequalities will lose their social significance.

6. Equality and Difference

Material inequality ceases to be a major preoccupation when it ceases to be the symbol of hierarchical stratification: material well-being is neither insulting nor impoverishing for others when it is not accompanied by invidious distinctions or power over other people's lives. Physical poverty is not humiliating when it proceeds from choosing to be satisfied with less and not from being relegated to the lower ranks of society.

The unwillingness of many contemporary Marxists to recognize these facts demonstrates to what extent their own cultural universe and value-system have been flattened by commodity relations; inequality for them signifies not merely that people are "different" but that they are "higher up" or "lower down," depending on whether they earn "more" or "less". It is, however, this one-dimensionality of values, lifestyles, and individual goals which has permitted the extension of commodity relations and wage-labor to all domains of human activity. Competition, resentment, and acquisitiveness in the name of equality or "social justice" are only possible in a socially homogeneous universe where differences are of a purely quantitative and hence measurable character. The categories of "more" and "less" presuppose a sociocultural *continuum* in which inequality is conceived only as an economic difference between inherently equal individuals.

This spurious definition of inherent equality is the cultural foundation of capitalism: it is what gave rise to, or at least made plausible, the monetary evaluation of all differences and their translation into income inequalities. Hence the fierce repression, associated with the rise of the bourgeoisie, of those minorities and cultural deviants who—by their attachment to the uniqueness and otherness of their values—threatened the one-dimensionality of the sociocultural system essential to the dominance of commodities. Hence the idea of universal compulsory education, which we now recognize, tends by its very uniformity to favor the most privileged. Hence the growing antipathy of the government to the claims of professional ethics—the tradition of principled autonomy which the members of different professions could invoke to refuse the sale or hire of their skills.

The meaning and the content of each activity have thus been suppressed and replaced by a monetary "compensation," that is to say, by a certain exchange value. Increasing the amount of this compensation becomes the overriding objective of all productive social activity: of "work." Work is thus emptied of all substance, reduced to a tribute measured by its duration, and purchased from the worker like any other commodity. It is our income which determines our worth, not our activity—which is stripped of all independent purpose. Alienation of labor makes money (purchasing power) into the principal aim of the individual.

This is what lies behind the unending pursuit of an ever-receding equality: those in each income category seek parity with those at the next level of income who, in turn, attempt to "catch up" with those above them. Beyond a certain level, increases in income are sought not for their own sake or for the additional consumption which they represent. Interestingly, they reflect above all the demand that society recognize us as having the same rights, the same standing, and social value we see attributed to others. In a society based on the unequal remuneration of jobs equally devoid of meaning, the demand for equality is the hidden source of the continuing escalation of consumer demand, dissatisfaction, and social competitiveness.

The stabilization of the level of consumption will thus re-

main impossible until:
- all socially necessary tasks receive equal social recognition (and rewards); and
- the possibility is given to everyone to actualize the infinite diversity of abilities, desires, and personal tastes through an unlimited variety of free individual and collective activities.

The reduction of the duration of socially necessary labor and the possibility of using one's free time in productive ways are the essential preconditions for the disappearance of commodity relations and competition. Different standards of living and lifestyles will cease to signify inequality when they are the result not of differences in income but of the diversity of pursuits by communities and individuals during their free time.

7. Social Self-Regulation and Regulation from Outside: Civil Society and the State

The rift between production and consumption, between work and "leisure," is the result of the destruction of autonomous human capabilities in favor of the capitalist division of labor. This rift enables the sphere of commodity relations to be perpetuated and indefinitely extended. Having been deprived of all possibility of control over the purpose or the character of labor, the realm of freedom becomes exclusively that of non-work periods. But since all creative or productive activity of any social consequence is nevertheless denied during "free" time, this freedom is itself reduced to a choice amongst objects of consumption and passive distractions.

The destruction of the autonomous capabilities of the worker does not therefore result *solely* from the fragmentation of work and the elimination of skills introduced by the "scientific organization of labor." It is not enough to attack the organization of labor. The destruction of autonomous capacities is carried out prior to the division of labor; it is accomplished by schooling.

The basic lessons taught in school are that there is a competent authority for every question and a specialist for every task; that the "all-sided" individual, whom Marx refers to as "integral" because his or her capabilities have been fully developed, can never be anything but a "dilettante" or a "dabbler." Schooling discourages independence and versatility in favor of graded "qualifications," which have the essential characteristic of having no use value for the person who acquires them, but only an exchange value in the marketplace. You can't do anything for yourself with what you learn in school. The only way to make use of the qualifications bestowed by schools is through the mediation of a third person, by trying to sell oneself on the "job market."

Schools do not teach us how to speak foreign languages (or even our own, for that matter), how to sing or use our hands and feet, how to eat properly, how to cope with the intricacies of bureaucratic institutions, how to look after children or take care of sick people. If people do not sing any more but buy millions of records to have professionals sing for them, if they don't know how to nourish themselves but pay doctors and the pharmaceutical industry to treat the symptoms of an improper diet, if they don't know how to raise children but only how to hire the services of childcare specialists "certified by the state," if they don't know how to repair a radio or fix a leaky faucet or take care of a strained ankle or cure a cold without drugs or grow a vegetable garden, etc., it is because the unacknowledged mission of the school is to provide industry, commerce, the established professions, and the state with workers, consumers, patients, and clients willing to accept the roles assigned to them.

The institutional function which has been passed on to the school is to perpetuate and confirm—not to counter or correct—the disintegrating, infantilizing, and deculturing action of society and the state. In an educative civil society—that is to say, one underpinned by a living culture—the school could not have the effects that it has or be what it is today. It is what it is because it participates in the general process whereby knowledge, culture, and autonomy are expelled from work, from life outside work and the space in which it is lived, from the relations between people and with nature, to be concentrated in specialized institutions where, inevitably, they become institutional specialities.

Unemployment, i.e., the inability to produce other than by working for someone else, is the final absurdity of a system based on regulation from outside.

The destruction of autonomous capabilities is thus to be

understood as part of a process, in part deliberately planned, tending to strengthen the domination of capital—or of the state which assumes its functions—over the worker not only as a worker but also as a consumer. By making it impossible for individuals to produce, within the extended family or the community, any of the things which they consume or aspire to, capitalism (and the state) forces them to satisfy the totality of their needs by commodity consumption (i.e., by the purchase of institutionally produced goods and services); at the same time, *capitalism reinforces its control over this consumption.*

This destruction of autonomous capabilities and the cultural uniformity which it brings about are necessarily associated with the destruction of civil society by the state. By "civil society" I mean the web of social relations that individuals establish amongst themselves within the context of groups or communities whose existence does not depend on the mediation or institutional authority of the state.[18] It includes all relations founded upon reciprocity and voluntarism, rather than on law or judicial obligation.

It includes, for example, the relations of cooperation and mutual aid which can arise in communities, neighborhoods, or among residents of the same building; the cohesion and solidarity of older working class areas; the voluntary associations and cooperatives created by people themselves in their common interest; the family relations and larger domestic communities; in short, the totality of exchanges and communications which constitute or once constituted the "life" of the neighborhood or small town.[19]

This whole web of self-regulating and noninstitutional social relations is dislocated by the social and territorial division of labor which accompanies industrialization. Rural depopulation destroys village communities, swells the suburbs and juxtaposes isolated individuals in dormitory cities whose physical design presents further obstacles to communication and personal exchange. The length of travel between home and work increases fatigue. And the crowding of cities, streets, and transit systems makes of us "all" that pure quality of anonymous humanity which, by its very density, constitutes an obstacle to the comfort and mobility of "each."

Work itself is suffered rather than accomplished, the workers being shaped by the machine served rather than making it serve themselves in the shaping of inanimate matter. This work blunts their faculties and leads to the atrophy of their capacity to produce for themselves.

Fatigue, lack of space, lack of time, and lack of neighborhood interactions contribute to the decline of mutual aid: commercialized services—eventually supplemented by public agencies, household appliances, and subsidized facilities—come to fulfill the roles previously assigned to parents, relatives, and neighbors.

This decline of civil society is everywhere accompanied by a reinforcement and an expansion of the institutional activities of the state. Disconnected individuals call on the state to compensate, by an ever-greater social presence, for the disappearance of their capabilities to help each other, to protect each other, to care for each other, and to raise their own children. The extension of institutional responsibility promotes further professionalization, specialization, and the subversion of all activities—hence accelerating the decline of civil society.

This displacement of civil society by the state corresponds, at the political level, to the replacement of self-regulation by regulation from outside. What has been said about natural selection applies equally here. Regulation from outside can indeed be more efficient than self-regulation: the concentration of production in large units, central planning (whether by corporate management or by the state), the fragmentation of work, and the resulting quasi-militarization of the workforce can be accompanied at least up to a certain point by increased efficiency.

Industrial concentration entails, however, an inevitable geographical concentration and specialization of functions. The result is that each geographical collectivity—neighborhood, town, city, region—no longer functions in relation to its own needs but produces to serve the totally abstract needs of faraway and anonymous users. No one consumes what he or she produces or produces what he or she consumes. The production of large specialized factories is necessarily regulated externally by the "market" and/or the state, which is to say by other large institutions (banks, brokerages, sales offices, administrative agencies) specializing in regulation from outside.[20]

The improvement in efficiency thus has as its counterpart a proliferation of bureaucracy which entails growing costs, rigidities, and slowdowns of its own; increases the centralization of power and the uniformity of the individual; and—beyond a certain point—leads inevitably to wastefulness, squandering of energy and resources, and eventually diminished efficiency. The withering away of civil society under the aegis of the state thus initiates the withering away of basic freedoms and the establishment of a more or less militarized social system in which the state runs everything. It is customary to call such societies "totalitarian" because in them the state has wholly supplanted civil society and has become a "total state."[21]

We have virtually reached this stage today. No social or cultural activity, no civic improvement or productive process can be initiated by those directly concerned without the intervention, authorization, regulation, or supervision of some "competent authority." No initiative can be taken from below without designating someone as "responsible"—responsible not to fellow citizens but before the law. No work can be done or carried out unless it is *assigned,* i.e., unless its character and purpose have been established by an institutional "employer." No voluntary association of individuals can be formed without having to give an institutional account of itself, without having the established leadership attempt to subsume its activities or circumscribe its objectives.

With needs determined by a series of institutions, professions, prescriptions, and rights, the citizen is invited to behave primarily as a consumer, a customer, a client who is legally entitled to a series of services, facilities, and forms of assistance. The citizen no longer consumes those goods and services which correspond to the autonomous needs which he or she feels, but those which correspond to the heteronomous needs attributed to him or her by the professional experts of specialized institutions.[22]

The divergences between contending political parties are mainly over the character and extent of institutional treatment to be meted out for institutionally defined needs. In politics, too, citizens are treated as consumers of policies devised and implemented for them by those "in charge": they can choose between political parties in the same way they can choose between different brands of detergent. Let an individual refuse this choice and he or she will be dismissed as "apathetic." Discouraged from doing anything by or for themselves, deterred from associating with others in order to create—according to their own preferences—their own way of working, of housing themselves, of producing, moving about, consuming, living, people are encouraged instead to seek new forms of assistance, "from above," to fill up the last spaces left open to their own initiative.

Against this fundamental tendency, the limited "self-management" of municipalities or factories is helpless to withstand or counteract the increasing hegemony of the state. What is required at the same time is that the size, functioning, and organization of communities and institutions be opened up to provide new spaces for free action which permit self-regulation to bear upon the *what* and not only the *how*.

Local self-management of centrally regulated units is an absurdity, or at least a mystification. Such "self-management" is necessarily instituted by the system or by the state itself, and hence has lost its autonomy even before it has gained it. It can in no way obviate or even significantly modify the hazards and constraints inherent in large systems, whose very scale and complexity require the coordination and external regulation of their various units.

Self-management is meaningless in a concentrated and specialized economy. Large cities which have specialized in the production of a single commodity, such as steel or tires, are dependent on business cycles and market fluctuations beyond their influence or control. Demands for local self-determination and/or worker management of factories are vacuous where big business corporations, or even worse, a single specialized subsidiary, are the sole employers and by far the main taxpayers.

Self-management necessarily entails social and economic units that are small enough and diverse enough to provide the community with outlets for a wide variety of human talents and capacities, with the basis for a rich diversity of human exchanges and interactions, and with the possibility of adjusting at least part of the production to local needs and preferences, thus ensuring a basic minimum of self-reliance.

In short, self-management presupposes tools capable of being self-managed. The creation of these tools is technically feasible. It is not a question of reverting to cottage industry, to the village economy, or the Middle Ages, but of subordinating industrial technologies to the continuing extension of individual and collective autonomy, instead of subordinating this autonomy to the continuing extension of industrial technologies.[23] In Illich's terms, "the value of the system of tools depends on its ability to integrate the outputs of heteronomous production with the spontaneous desires and personal needs of the people."

The redefinition and redistribution of the system of tools evidently presupposes a restructuring of societal institutions and of the state. There can be no question of abolishing the latter by a single stroke, but only of making it wither away through the expansion of civil society.[24]

Against the centralizing and totalitarian tendencies of both the classical Right and the orthodox Left, ecology embodies the revolt of civil society and the movement for its reconstruction.

8. Seven Theses by Way of Conclusion

A number of conclusions can be drawn from the partial analyses which make up this essay. I shall try to state them succinctly here, as theses, and then illustrate them in the form of a utopia for modern society.

1. The causes of the current crisis of capitalism are the over-development of productive capacities and the destructiveness of the technologies they are based on; this over-development and destructiveness aggravates existing scarcities while generating new ones. The crisis cannot be overcome except by a new mode of production which, breaking with current economic rationality, is based on the careful stewardship of renewable resources and the decreasing consumption of energy and raw materials.

2. The overcoming of current rationality and the reduction of material consumption can be brought about by technofascist

central regulation as well as by convivial self-management. Technofascism cannot be prevented except by the expansion of civil society, which depends in turn on the expansion of tools and technologies which foster the sovereignty of the individual and the community.

3. The connection between "more" and "better" has been broken. "Better" may now mean doing with less. It is possible to live better by working and consuming less, provided we produce more durable things as well as things which do not destroy the environment or create insurmountable scarcities once everyone has access to them. *Social production should be reserved for those things which remain useful to each when distributed to all—and vice versa.*

4. Poverty in wealthy countries is caused not by insufficient production but by the kinds of goods produced, the methods used to produce them, and their inequitable distribution. Poverty will not be eradicated until there is an end to the *social* production of scarce luxuries, that is to say, of goods which are reserved and exclusive by nature.[25] *Only that which neither privileges nor demeans anyone deserves to be produced socially.*

5. Unemployment in wealthy societies reflects the decreasing amount of socially necessary labor time. It demonstrates that everyone could work less provided everyone worked. The equal social recognition and remuneration of all socially necessary work is the essential condition for both the elimination of poverty and the distribution of work amongst all those capable of it.

6. Once social labor is limited to that required for socially necessary production, the reduction of working hours can be accompanied by the expansion of self directed and freely chosen activities. Over and above the essentials guaranteed by social production, people will be able to use their free time to produce, individually or collectively, whatever else seems appropriate to them. The production of an unlimited variety of goods and services by neighborhood cooperatives and in neighborhood facilities will ensure the expansion of the realm of freedom and the decline of commodity relations—the expansion of civil society and the withering away of the state.

7. The uniformity of consumption patterns and of lifestyles which characterizes present society will disappear with the disappearance of social inequality. Individuals and communities will distinguish themselves and diversify their patterns of living beyond anything conceivable today. These differences will, however, be the result of the different uses to which they put their time and resources, and not of unequal access to power and social rewards. The development of autonomous activities during the free time available to everyone shall be the only source of distinction and of wealth.

To illustrate these theses, I shall describe one of several possible utopias. The conclusions stated above could of course be given a different expression from the one suggested here: its only function is to liberate the imagination as to the possibilities for change.

.   .   .   .   .

When they woke up that morning, the citizens asked themselves what new turmoil awaited them. After the elections, but during the period of transition to the new administration, a number of factories and enterprises had been taken over by the workers. The young unemployed, who for the previous two years had been occupying abandoned plants in order to engage in "wildcat production" of various socially useful products, were now joined by a growing number of students, older workers who had been laid off recently, and retired people. In many places, empty buildings were being transformed into communes, production cooperatives, or "alternative schools." In the schools themselves, the older pupils were taking the lead in practicing skills for self-reliance and, with or without the collaboration of the teachers, establishing hydroponic gardens and facilities for raising fish and rabbits; in addition, students were beginning to install equipment for woodworking, metalworking, and other crafts which had for a long time been neglected or relegated to marginal institutions.

The day after the new government came into office, those who set out for work found a surprise awaiting them: during the night, in most of the larger cities, white lines had been painted on all the major thoroughfares. Henceforth these would have a corridor reserved for buses, while on the sidestreets similar corridors were set aside for bicyclists and motorcyclists. At the major points of entry to each city, hundreds of bicycles and mopeds were assembled for use by the public, and long lines of police cars and army vans supplemented the buses. On this morning, no tickets were being sold or required on the buses or suburban trains.

At noon, the government announced that it had reached the decision to institute free public transportation throughout the country, and to phase out, over the next twelve months, the use of private automobiles in the most congested urban areas. Seven hundred new tramway lines would be created or reopened in the major metropolitan centers, and twenty-six thousand new buses would be added to city fleets during the course of the year. The government also announced the immediate elimination of sales tax on bicycles and small motorbikes, thus reducing their purchase price by twenty per cent.

That evening, the President of the Republic and the Prime Minister went on nationwide television to explain the larger design behind these measures. Since 1972, the President said, the GNP per person in France has reached a level close to that of the United States—the difference varying between five and twelve per cent according to the fluctuations in the value of the franc, which has been notoriously undervalued. "Indeed, my fellow citizens," the President concluded, "we have nearly caught up with the U.S. But," he added soberly, "this is not something to be proud of."

The President reminded his listeners of the period, not so distant, when the standard of living of Americans seemed an impossible dream to French men and women. Only ten years ago, he recalled, liberal politicians were saying that once the French worker began earning

American wages, that would be the end of revolutionary protests and anticapitalist movements. They had been, however, profoundly mistaken. A large proportion of French workers and employees were now receiving salaries comparable to those being paid in the U.S. without this having diminished the level of radical activism. "On the contrary. For in France, as in the United States, the people find themselves having to pay more and more to maintain an increasingly dubious kind of well-being. We are experiencing increasing costs for decreasing satisfactions. Economic growth has brought us neither greater equity nor greater social harmony and appreciation of life. I believe we have followed the wrong path and must now seek a new course." Consequently, the government had developed a program for "an alternative pattern of growth, based on an alternative economy and alternative institutions." The philosophy underlying this program, the President stated, could be summed up in three basic points:

1. "We shall work less." Until now, the purpose of economic activity was to amass capital in order to increase production and sales, and to create profits which, reinvested, would permit the accumulation of more capital, and so on. But this process must inevitably reach an impasse. Beyond a certain point, it could not continue unless it destroyed the surplus which it had created. "We have reached that point today," the President said. "It is, in fact, only by wasting our labor and our resources that we have managed in the past to create a semblance of the full employment of people and productive capacities."

In the future, therefore, it was necessary to consider working less, more effectively, and in new ways. He said that the Prime Minister would spell out the details of proposed measures for change in this direction. Without going into them, the President nevertheless stated that they would give substance to the following principle: "Every individual will, as a matter of right, be entitled to the satisfaction of his or her needs, regardless of whether or not he or she has a job." He argued that once the productive machinery reaches the level of technical efficiency where a fraction of the available workforce can supply the needs of the entire population, it is no longer possible to make the right to a full income dependent on having a full-time job. "We have earned," the President concluded, "the right to free work and to free time."

2. "We must consume better." Until now, products had been designed to produce the greatest profit for the firms selling them. "Henceforth," the President said, "they will be designed to produce the greatest satisfaction for those who use them as well as for those who produce them."

To this end, the dominant firms in each sector would become the property of society. The task of the great firms would be to produce, in each area, a restricted number of standardized products, of equal quality and in sufficient amounts, to satisfy the needs of all. The design of these products would be based on four fundamental criteria: durability, ease of repair, pleasantness of manufacture, and absence of polluting effects.

The durability of products, expressed in hours of use, would be required to appear alongside the price. "We foresee a very strong foreign demand for these products," the President added, "for they will be unique in the world."

3. "We must re-integrate culture into the everyday life of all." Until now, the extension of education had gone hand in hand with that of generalized incompetence.

Thus, said the President, we unlearned how to raise our own children, how to cook our own meals and make our own music. Paid technicians now provide our food, our music, and our ideas in prepackaged form. "We have reached the point," the President remarked, "where parents consider that only state-certified professionals are qualified to raise their children adequately." Having earned the right to leisure, we appoint professional buffoons to fill our emptiness with electronic entertainment, and content ourselves with complaining about the poor quality of the goods and services we consume.

It had become urgent, the President said, for individuals and communities to regain control over the organization of their existence, over their relationships and their environment. "The recovery and extension of individual and social autonomy is the only method of avoiding the dictatorship of the state."

The President then turned to the Prime Minister for a statement of the new program. The latter began by reading a list of twenty-nine enterprises and corporations whose socialization would be sought in the National Assembly. More than half belonged to the consumer goods sector, in order to be able to give immediate application to the principles of "working less" and "consuming better."

To translate these principles into practice, the Prime Minister said it was necessary to rely on the workers themselves. They would be free to hold general assemblies and set up specialized groups, following the system devised by the workers of Lip, where planning is done in specialized committees, but decisions are taken by the general assembly. The workers should allow themselves a month, the Prime Minister estimated, to define, with the assistance of outside advisers and consumer groups, a reduced range of product models and new sets of quality standards and production targets. New management systems had already been devised by a semi-clandestine group of Ministry of Finance officials.

During this first month, said the Prime Minister, production work should be done only in the afternoons, the mornings being reserved for collective discussion. The workers should set as their goal the organizing of the productive process to meet the demands for essential goods, while at the same time reducing their average worktime to twenty-four hours a week. The number of workers would evidently have to be increased. There would, he promised, be no shortage of women and men ready to take these jobs.

The Prime Minister further remarked that the workers would be free to organize themselves in such a way that each individual could, for certain periods, work more or less than the standard twenty-four hours for the same firm. They would be free to have two or three part-time jobs, or, for example, to work on construction during the spring and in agriculture towards the end of the summer—in short, to learn and practice a variety of skills and occupations. To facilitate this process, the workers themselves would be helped to set up a system of job exchanges, taking into account that the 24-hour week, and the monthly salary of 2000F ($500) to which they would be entitled, should be regarded as an average.

Two people, said the Prime Minister, should be able to live quite comfortably on 2000F a month, considering the range of collective services and facilities which would be available to them. But no one need feel restrained by this: "Luxuries will not be prohibited. But they must be obtained by additional work." As examples, the Prime Minister cited the following: a secondary residence or summer cottage represented about three thousand hours of labor. Anyone seeking to acquire one would work, in addition to the twenty-four hours a week, three thousand hours in the building and construction sector, of which at least a thousand hours would need to be completed before a loan could be raised. Other objects classified as non-necessities, such as private automobiles (which represented about six hundred hours of labor), could be acquired in the same fashion. "Money itself will no longer confer any rights," the Prime Minister stated. "We must learn to determine the prices of things in working hours." This labor-cost, he added, would rapidly decline. Thus the individual with some do-it-yourself skills would soon be able to acquire, for only five hundred hours of additional work, all the elements needed to assemble his or her own house, which should not take more than fifteen hundred hours to put up.

The government's economic aim, the Prime Minister stated, was to gradually eliminate commodity production and exchange by decentralizing and scaling down production units in such a way that each community was able to meet at least half of its needs. The source of the waste and frustration of modern life, the Prime Minister noted, was that "no one consumes what he or she produces and no one produces what he or she consumes."

As a first step in the new direction, the government had negotiated with the bicycle industry an immediate thirty per cent increase in production, but with at least half of all the bicycles and motorcycles being provided as kits to be put together by the users themselves. Detailed instruction sheets had been printed up, and assembly shops with all the necessary tools would be installed without delay in town halls, schools, police stations, army barracks, and in parks and parking lots....

The Prime Minister voiced the hope that in the future local communities would develop this kind of initiative themselves: each neighborhood, each town, indeed each apartment block, should set up studios and workshops for free creative work and production; places where, during their free time, people could produce whatever they wished thanks to the increasingly sophisticated array of tools which they would find at their disposal (including stereo equipment or closed-circuit television). The 24-hour week and the fact that income would no longer depend on holding a job would permit people to organize so as to create neighborhood services (caring for children, helping the old and the sick, teaching each other new skills) on a cooperative or mutual-aid basis, and to install convenient neighborhood facilities and equipment. "Stop asking, whenever you have a problem, 'What is the government doing about it?'" the Prime Minister exclaimed. "The government's vocation is to abdicate into the hands of the people."

The cornerstone of the new society, the Prime Minister continued, was the rethinking of education. It was essential that, as part of their schooling, all young people learn to cultivate the soil, to work with metal, wood, fabrics, and stone, and that they learn history, science, mathematics, and literature in conjunction with these activities.

After completing compulsory education, the Prime Minister went on, each individual would be required to put in twenty hours of work each week (for which he or she would earn a full salary), in addition to continuing with whatever studies or training he or she desired. The required social labor would be done in one or more of the four main sectors: agriculture; mining and steelworks; construction, public works, and public hygiene; care of the sick, of the aged, and of children.

The Prime Minister specified that no student-worker would, however, have to perform the most disagreeable jobs, such as collecting garbage, being a nurse's aide, or doing maintenance work, for more than three months at a time. Conversely, everyone up to the age of forty-five would be expected to perform these tasks for an average of twelve days a year (12 days a year could mean one day per month or one hour per week). "There will be neither nabobs nor pariahs in this country any more," he remarked. In a matter of two years, six hundred multi-disciplinary centers of self-learning and self-teaching, open day and night, would be put within easy reach of everyone, even of people living in rural areas, so that no one would be imprisoned in a menial occupation against his or her choice.

The student-workers would also be expected, during their last year of work-education, to organize themselves into small autonomous groups to design and carry out an original initiative of some kind, which would be discussed beforehand with the local community. The Prime Minister expressed the hope that many of these initiatives would seek to give new life to the declining rural regions of France, and serve to reintroduce agricultural practices more in harmony with the ecosystem. Many people, he said, were unduly worried by the fact that France depends on foreign sources for gasoline and industrial fuel, when it was far more serious to be dependent on American soybean meal to raise beef, or on petrochemical fertilizers to grow grains and vegetables.

"Defending our territory," the Prime Minister said, "requires first of all that we occupy it. National sovereignty depends first of all on our capacity to grow our own food." For this reason the government would do everything possible to encourage a hundred thousand people a year to establish themselves in the depopulated regions of the country, and to reintroduce and improve organic farming methods and other "soft" technologies. All necessary scientific and technical assistance would be provided free for five years to newly established rural communities. This would do more to overcome world hunger, he added, than the export of nuclear power stations or insecticide factories.

The Prime Minister concluded by saying that, in order to encourage the exercise of imagination and the greater exchange of ideas, no television programs would be broadcast on Fridays and Saturdays.

NOTES

1. See the chapter below, "Reinventing the Future."

2. See the chapter below, "The Social Ideology of the Motorcar."

3. On the various levels of counterproductivity, see Ivan Illich: *Medical Nemesis: The Expropriation of Health* (New York: Pantheon, 1976); and Jean-Pierre Dupuy and Jean Robert: *La Trahison de l'opulence* (Paris: Presses Universitaires de France, 1976).

4. Cf., in Ivan Illich: *Tools for Conviviality* (New York: Harper and Row, 1973), p. 122, the following remarks, no doubt written with the Club of Rome in mind:

> A well-organized elite, vocally promulgating an antigrowth orthodoxy, is indeed conceivable. It is probably now forming. But such a programmatic antigrowth elite would be highly undesirable. By pushing people to accept limits to industrial output without questioning the basic industrial structure of modern society, it would inevitably provide more power to the growth-optimizing bureaucrats and become their pawn. One of the first results of transition toward a stable-state industrial economy would be the development of a labor-intensive, highly disciplined, and growing subsector of production that would control people by giving them jobs. Such a stabilized production of highly rationalized and standardized goods and services would be—if this were possible—even further away from convivial production than the industrial-growth society we have now.

5. I borrow this expression from Illich (*Tools for Conviviality*, p. 23), who defines it as follows:

> For a hundred years we have tried to make machines work for men and to school men for life in their service. Now it turns out that machines do no 'work' and that people cannot be schooled for life at the service of machines. The hypothesis on which the experiment was built must be discarded. The hypothesis was that machines can replace slaves. The evidence shows that, used for this purpose, machines enslave men. Neither a dictatorial proletariat nor a leisured mass can escape the dominion of constantly expanding industrial tools.
>
> The crisis can be solved only if we learn to invert the present deep structure of tools; if we give people tools that guarantee their right to work with high, independent efficiency, thus simultaneously eliminating the need for either slaves or masters and enhancing each person's range of freedom.

6. And not just with that of growth-oriented capitalism. The end of the growth-oriented form does not necessarily sound the death-knell of the capitalist system: capitalism has already survived long periods of stagnation and crisis (1874-1893, 1914-1939). It requires the accumulation of capital; but when this becomes structurally impossible, far from crumbling, it works to make it possible again. Which may entail the massive destruction of capital and/or wars.

7. Cf., André Gorz, ed., *Division of Labour* (London: Harvester, 1977), and see the chapter below, "From Nuclear Electricity to Electric Fascism."

8. See the chapter below, "Science and Class: The Case of Medicine."

9. See the chapter below, "From Nuclear Electricity to Electric Fascism."

10. See the chapter below, "Socialism or Ecofascism."

11. See the chapter above, "Two Kinds of Ecology."

12. Based on a survey conducted by SOFRES for *Elle* magazine, March 1974.

13. Marshall Sahlins, *Stone Age Economics* (Chicago: Aldine, 1972).

14. Socialists traditionally place exclusive emphasis on this form of appropriation, as though it could explain all social and economic ills. It must be carefully pointed out, therefore, that not all (private) appropriation is detrimental, and that private property is neither the only nor the most important cause of poverty in industrialized societies. It causes poverty only when the limited resources monopolized by the rich would be available in sufficient quantities were no one allowed to own more than his or her share.

Appropriation is not detrimental when resources—e.g., land, water, fishing and hunting rounds, etc.—are overabundant and, for all practical purposes, unlimited. Appropriation is detrimental, but not the sole cause of poverty, when resources are so short that there would not be enough for everyone even if they were equally distributed.

Appropriation is both detrimental and the cause of poverty when a dominant minority—the ruling class or caste—opposes the equitable distribution of a vital resource of which there would be enough for all if the wealthy did not have more than their share; they use their control over a vital resource—e.g., land or water—to subject the rest of the people to their economic, social, and political control. (Author's footnote to the English edition.)

15. It can also be done, as in the Soviet Union or China, by a political attribution of access rights.

16. On this point, see the chapter below, "Socialism or Ecofascism."

17. *Tools for Conviviality, op. cit.* See also the chapter below, "Reinventing the Future."

18. On this issue, see the excellent work by Pierre Rosanvallon, *L'Age de l'autogestion* (Paris: Le Seuil, coll. "Politique," 1976).

19. For a comparison of the older working class neighborhood with the modern comforts of newer highrise developments, see the interview with a worker from the Batignolles rehoused in 1971, in *Les Temps Modernes*, no. 314-315 (September-October 1972), pp. 616-625. This document, of an exceptional quality, was part of a survey carried out by students at the Nantes School of Architecture.

20. See the chapter below, "From Nuclear Electricity to Electric Fascism."

21. It was Nazism—National Socialism—which first proclaimed itself *der totale Staat.*

22. This is an idea taken from Illich, who develops it in detail in *Disabling Professions* (London: Marion Boyars, 1977). William Klein illustrates a similar concept in his film *The Model Couple.*

23. This is essentially what Illich is suggesting in the latter parts of *Medical Nemesis, op. cit.,* especially in the section entitled "Specific Counterproductivity," and in *Tools for Conviviality,* where he picks up and expands the idea of synergy between autonomous and heteronomous production. This synergy occurs when industrial products (such as bicycles, telephones, transistor radios, video-cassettes, etc.) facilitate the development of autonomous activities rather than obstructing them. [This discussion is considerably expanded in the French edition of the book; see *Némésis médicale,* Le Seuil, 1975, section 3— translator's note.]

24. On this point see Pierre Rosanvallon, *op. cit.,* as well as the annex to *Deuxième Retour de Chine,* by Claudie and Jacques Broyelle and Evelyne Tschirhart (Paris: Le Seuil, 1977).

25. Production is said to be "social" when it is carried out by wage labor at the behest of an institution (whether a business enterprise or an administrative agency). A servant's work is thus not social, although it earns him or her a wage; nor are the products which workers can produce on their own using the tools of "their" workplace.

Chapter II

Ecology and Society

1. Reinventing the Future

"A certain kind of growth is drawing to a close. Together we must invent another kind." These are the words of Valery Giscard d'Estaing; they could have been the words of any one of his opponents. But what kind of other growth? To what end? To accomplish what? A collateral issue: must we really have growth and is there no salvation without it?

But suppose it were the other way around. Suppose there were no salvation even in growth. Suppose that—except at the price of a complete overthrow of current institutions, methods, and behavior—growth brought not the "best" that it promises, but more and more unbearable frustration and harm, more and more formidable constraints. Do we need to change growth or to change what is produced, the method of production, the definition of needs, the method of satisfying them; in short, the mode of production and the way of life?

Two very different books attack these questions head on: *Tools for Conviviality* by Ivan Illich, and *L'Anti-Économique* by Jacques Attali and Marc Guillaume. Illich is a subversive Catholic who is examining industrial societies from a distance of several centuries. Jacques Attali and Marc Guillaume are economics professors at the Ecole Polytechnique who show to what extent the so-called "science of economics" is steeped in ideological assumptions, political choices, and indefensible anthropological statements, and to what extent basic theory must be reworked. In spite of the profound differences in their purposes and their styles, the two books agree on a number of essential points, beginning with these:

1) "The argument that says growth reduces inequalities is an intellectual fraud with no basis." (Attali & Guillaume)

2) "Many needs are created and maintained by the system"; it is therefore defective logic to justify the system by the fact that "it is the best way to assure the satisfaction of the needs it creates." (Attali & Guillaume)

Let's start with the first point. In 1962, the richest 10% of the French population had an income seventy-six times (76 times!) higher than the poorest 10%. In comparison, this ratio of inequality was 10 for Czechoslovakia, 15 for Great Britain, 20 to 25 for West Germany, and 29 for the United States. Ten years later, the French industrial production had doubled; but the ratio of inequality was practically unchanged, and it still amounted to 29 in the United States.

Furthermore, in France, as in the United States, the bulk (more than half) of all goods and services were and are produced for the most affluent 20% of the population. In other words, the affluence of the rich and the poverty of the poor have remained the same.

I hear the instant objection to this: "But the poor are better off than they were ten years ago." Or, "they consume more, they are less poor." Mistake. Double mistake. Because:

1) If it is true that the poor consume more goods and services, it doesn't follow that they live better.

2) Even assuming they live better, it doesn't follow that they are less poor.

Let's look at these two points more closely:

1) Consuming more, that is, having control over a larger quantity of goods and services, doesn't necessarily mean that things are better. It could mean that henceforth it will be necessary to pay for what used to be free. Or that it is necessary to spend much more (in constant dollars) to make up one way or another for the general deterioration of the environment. Are people living better when they are consuming an increasing amount of individual and mass transportation in commuting between their workplaces and their bedroom suburbs—which are getting farther and farther away? Are they living better because every five or six years they have to replace linens which once lasted more than a generation? Or because in place of tap water that has become disagreeable they buy more and more so-called mineral water? Are they living better because they consume more fuel to heat dwellings that are increasingly poorly insulated? Are they less poor because instead of hanging out at the corner cafe and the local moviehouse—both on their way out—they can buy a TV set and a car which offer them imaginary and solitary escape from the concrete deserts in which they live?

It's been quite a while since economists like Ezra Mishan[1] established that when the damage growth causes (injuries, pollution, breakdown of personal relationships) is taken into account, growth "signals further deterioration, and not improvement"; "it's cost is higher than the advantages it bestows." (Attali & Guillaume)

Or, as Illich writes, "the growth addicts are ready to pay an every increasing price for an ever decreasing enjoyment." The enormous spread of fast cars has effectively increased distances even more rapidly than the speed of the vehicles, and has obliged everyone to devote more time and money, space and energy, to traveling. "The speed industry is wrestling with the other industries to see who can strip man of the bit of humanity left to him...We cannot assume that growth aims at increasing general well-being. The defense that growth can be reoriented is not admissible unless the reorientation is radical." (Attali & Guillaume)

2) You may object, of course, by pointing out that electrical appliances have been "democratized"—they are no longer the privilege of the elite, as they were 40 years ago. And the same

goes for the consumption of meat, canned goods, cars, vacations...But does that mean that the workers, for example, are less poor? Put the question to older workers. They will tell you that in 1936, on their two weeks paid vacation a man and wife (or woman and husband) could go on a journey on bicycles, get room and board at a hotel for two weeks, and have some money left over when they came home.[2] Today, to earn enough for their vacation of staying at a hotel and traveling by car, both the man and woman must work and save. There is no time anymore for cooking and shopping, so they need a fridge, canned goods, fast food, and overtime to pay for it all. Is that living better? Is that the "quality of life" the household appliances bring?

Here is a comment from a reader of *France Nouvelle*[3]: "Primarily it's a matter of free time, of time to live...Let's fight for a five or six hour workday and the electric gadgets can go into a museum...What trouble is it to do laundry for four people when you can get home at four o'clock? Where's the bother in washing dishes for eight people when everyone in the family takes a turn?"

All the same, you may say, it's a fact that working class people have all the modern conveniences once the exclusive property of the middle class; therefore they are less poor. But wait a minute. Less poor than who? Than Indian or Algerian poor? Than the workers of 50 years ago? The comparison is completely abstract. Poverty is not an objective and measurable fact (which makes it different from destitution or starvation). It is a difference, an inequality, an inablility to acquire what society defines as "good." To be poor is to be excluded from the dominant lifestyle, and this lifestyle is never that of the majority, but of the 20% of the population whose privileged and ostentatious consumption dictates the style. In a society where everyone is poor, no one is. What makes poor people poor is that they are less well-off in relation to the sociocultural standard that directs and stimulates desires.

A person is poor in Peru when she or he has to go barefoot, in China when she or he has no bicycle, in France when she or he can't pay for an automobile. In the 1930s people were poor when they couldn't buy a radio, in the 1960s they were poor if they had to deny themselves a TV set, in the 1970s they become poor by not having a color TV. As Illich says, "poverty modernizes itself. Its financial threshold keeps rising because new industrial products are presented as basic necessities, even while they remain out of reach for most people." The masses of people "pay an ever increasing price for being increasingly under-privileged."

In effect, as soon as the mass consumer gets any product it is devalued. Sometimes, as in the case of the automobile, it is devalued by the simple fact that most people use it. The car loses its use value and becomes an impediment to travel and to the access people have to each other. The privileged minority then turns to new luxury transportation (special trains, airplanes, taxis, hired cars). Sometimes, even though the popular product hasn't lost any of its use value, the producer devalues it by introducing a "better" product which is available only to a minority and which, presented as the new definition of "well-being," will maintain inequality. "Innovation feeds the fantasy that what is new is better." It "creates more needs than it satisfies" and deepens frustration. "The level of frustration grows much faster than the level of production." (Illich) Because "if what is new is better, what is old is no good...The logic of 'ever better' replaces the logic of the simple good as the guiding principle of action."

In short, as Attali and Guillaume also show, following Baudrillard, maintenance of inequality propels economic growth. "The arrival of a new product on the market and its purchase by the rich gnaws at the poor until they can have it...Thus the dynamic between the social classes plays right into the producers' hands, and does not improve things at all. This dynamic at least partly explains how demand sustains growth."

Goods, to sum up, are no longer desired and bought for their use value, but for their "function as symbols of status, escape, and display." The individual is "brought up, educated" to want them. The social environment "imposes" this method of self-expression and affirmation by denying the individual "the pos-sibility of personal fulfillment in his or her work," by "diverting his or her desire into a desire to consume." (Attali & Guillaume)

On this point, however, Illich's analysis goes deeper than the Baudrillard analysis that inspires Attali. What is it, Illich asks, that allows needs and desires to be "diverted" into desire to consume? Answer: the fact that for the satisfaction of *all* needs a person is first of all reduced to dependence on giant institutions and tools that are out of the individual's control and grasp. Even for the (purified) air he or she breathes, for the (treated or bottled) water he or she drinks, for the sun (which the tourist industry sells), a person depends on the *mega-tools* of bureaucratic and commercial *mega-institutions* and is reduced to only being their client—submissive, standardized, powerless, exploited, and always dissatisfied.

Made passive, the person asks only that the mega-institutions which provide the goods take charge of his or her needs more completely and "better." Illich calls this being subject to a "radical monopoly." "The establishment of a radical monopoly happens when people give up their native ability to do what they can for themselves and for each other, in exchange for something 'better' that can be done for them only by a major tool... Radical monopoly imposes compulsory consumption." That is, it transforms the individual into a passive consumer "because it is enforced by means of the imposed consumption of a standard product that only large institutions can provide." In the end, even "basic needs can't be satisfied outside of the marketplace."[4]

This kind of analysis is perfectly acceptable and usable for Marxists. What Illich is describing is nothing but the extension of market relations into all areas of personal and social life, and their domination by industrial, banking, and government monopolies. What he is denouncing are simply the capitalist relations of production—which are supported by capitalism's division of labor. This means work suffers a division both technically (tasks are fragmented) and socially (there is a hierarchy of skills, pay and power), which separates the producers from the means of production and from the products so as to subject the workers more effectively to the demands of Capital (to the rules set by management and the speed of the machines). The more gigantic the means of production, the better they guarantee this subjection. For this makes them less controllable and usable by the workers who are subjected to them and by the community (town or district) where they are set up.

It's very important to avoid saying that this gigantism of

"tools" and the division of labor it enforces are the inevitable consequence of the "development of the forces of production," and of technico-scientific progress. The contrary has been decisively demonstrated by an American scholar; and intelligent employers as well as somewhat imaginative scientists know that gigantism is not a technical necessity but a political choice.[5] Middle-sized units of production (not more than 500 workers) are more efficient, more creative with innovations and inventions (the OECD furnishes statistical proof of this), and more economical (fewer rejects, external diseconomies, pollutants, etc.).

The reasons why capitalism is allergic to medium-sized units are essentially political. The smaller plants are too easy for the workers to take in hand, as a recent series of French strikes has shown (Jaeger, Lip, Cérizay, etc.); and for the employer they have this additional disadvantage: unlike larger-sized units, they don't allow the employer to control local politics and the local job market.

Far from requiring gigantism, science and technology have given birth to outsize tools because capital demands these tools and refuses any others. Windmills, as the great historian Marc Bloch has shown, lost out only because, since the wind is everywhere, they couldn't be monopolized. In short, as Illich says in unexpectedly Marxist language, "the structure of the forces of production shapes the social relations," precisely because it has itself been shaped to guarantee the domination of capital over labor.

From this point on, Illich's and Attali's opinions link up again concerning what can and cannot be a socialist society.

For Illich, the possibility of adapting anti-convivial tools which manipulate and enslave the individual to a socialist society is extremely unlikely. Public ownership of the means of production under the tutelage of a planning board could not transform the anti-human structure of the tool. "As long as the Ford Motor Company can be condemned simply because it makes Ford rich, the illusion is bolstered that the same factory [i.e., building cars on an assembly line] could make the public rich." Now, "the concept of ownership cannot be applied to a tool that cannot be controlled."[6] It cannot be applied to *mega-tools*, whose control requires bureaucratized and hierarchical administrative machinery which crushes people and generates centralized power. "Limited resources can be used to provide millions of viewers with the color image of one performer or to provide many people with free access to produce and distribute their own programs. In the first case, technology will be used for the further promotion of the specialist who is controlled by bureaucrats...But science can also be used to simplify tools, and to enable the layperson to shape his or her immediate environment to his or her taste."[7]

In the same way Attali and Guillaume write:

> Ought we to give power to those who don't have it, or try to take it away from everyone?...The idea of self-management seems at the moment the only new proposal available. But that's not enough to establish an overall model. A first step toward democracy and non-power, self-management could easily slide in the direction of the current industrial system and its contradictions. Self-managing workers at General Motors wouldn't be less of a pressure group favoring the development of the automobile than the current financial lobbies...Self-regulated enterprises that have autonomy but whose fundamental social relations are unchanged will lead to a kind of workers' capitalism to which the present mode of production will be very amenable.

But it is not just the mode of production, it's the entire economic logic that needs to be changed. And from this perspective, "what is essential is not to define a new coherent political scheme, but to suggest a new imaginative attitude, one that will be radical and subversive, by which alone we will be able to change the logic of our development."

This "proposal of a rupture, of a disassembly of the economic system, can only take place outside of monopolistic capitalism and bureaucratic socialism. It can also only take place outside any reference to an existing model which would inevitably be compromising. That is, it must above all call into question *the legitimacy of any power*, and reject both capitalist exploitation and totalitarian alienation."

Like Illich, Attali and Guillaume thus refuse ready-made top-down solutions. The point is not to govern people and economic processes better, but to allow everyone to take their lives into their own hands and change them, to free themselves from "external powers" and "external goals" [Marx] while establishing a radically new economy—an economy that will function "through a different standard of personal behavior [putting aside selfishness, ownership, and power] and not simply alternative procedures." (Attali & Guillaume)

The "different standard of behavior" cannot be the result of manipulation or teaching but can only be achieved by basic change and the liberating discovery [encouraged by the crises and dilemmas of the industrialized world] that it is possible to make more with less, to create "more happiness for everyone with less affluence." (Illich) Limitation of growth is not a goal in itself; it is of no interest if it is advocated and imposed by a "new organized elite" whose whole program is to be anti-growth. On the contrary, the formation of such an elite "is the industrial antidote to the revolutionary imagination. By urging the people to accept a limitation of industrial production without questioning the basic structure of industrial society, one is necessarily giving more power to the bureaucrats who optimize growth, and one becomes oneself hostage to that power."

In short, we must refuse to allow the capitalist managers to take over a critique of growth, a critique that only makes sense—and has revolutionary meaning—in reference to a "total social change," to a "change in the mechanisms that have shaped needs as they are today." But "everything seems to conspire everywhere to block, forbid, and distort the necessary imaginative subversiveness—and even simple verbal escape, except in the most standard formulations. The adoption of a socialist vocabulary by capitalist societies perverts its meaning. Ideological confusion is further aggravated by political debate that limits the options to a simplistic choice between market economy and centrally planned economy, when neither has ever been operative anywhere. Keeping this ideological block going, one runs the risk of making impossible any step towards a different future." (Attali & Guillaume) A future "where everyone would have a voice and share it, where no one would be able to limit anyone else's creativity, where everyone would be able to change his [or her] life." (Illich)

4 March 1974

2. Affluence Dooms Itself

There's no need to wait for it any longer—the great crisis has already begun. If you are having trouble recognizing it, that's because it is not taking the same form as last time.

This time the first sign is not the breakdown of capitalist production in the cities. It is primarily the collapse of all that gave it meaning: the bond between "more" and "better" is broken. Already the underside of the growth of production can be seen in the even greater growth of the damage it causes. People are living worse while consuming more. Growth is causing more scarcity than it relieves.

If you don't believe me, look around you. And read, for example, *L'Utopie ou la Mort*, by René Dumont.[1] Did you know that, among other things, the paper, furniture, and lumber companies that—with the blessing of Brazilian technocrats—are right now cutting down the Amazon forest are attacking the regenerative source of a quarter of the oxygen in our planet's air? Did you know that in big cities oxygen is already in such short supply that the Tokyo cops, so as not to be asphyxiated at the intersections, use "oxygen fountains" where they go to breathe at regular intervals? Or that in Los Angeles on some days people are advised not to move around too much so as to economize the bit of oxygen the cars and trucks have left for their lungs?

Did you know that Holland imports drinking water from Norway, the United States imports it from Canada, and that the city of San Francisco was thinking of hauling it down from the polar cap in icebergs? Did you know that, according to Cousteau, half of the marine life he filmed in 1956 had disappeared by 1963 (and what is left today)? Or that, according to the Soviet (Russian) Kasymov, the way things are going, by the end of the century the Caspian sea will be a stretch of water as pestilential, murky, and dead as Lake Erie is already?

Why? Because, for the market based production system that is dominant in Eastern as well as in Western Europe, that which has no price has no value. "That's no problem," exclaim the neoliberal economists, "we'll just put a price on those things that don't yet have one: air, water, light, and, of course, human life." For not even that is spared.

Did you know that one French laborer in six will be disabled during his or her working life? That all riveters and caulkers engaged in naval construction, all heavy-duty truck drivers, 45% of all ironworkers, and almost all steelworkers suffer from partial deafness? And that the newest chemical and petrochemical plants make this industry the one most dangerous to its workers' health?

Then, dear neoliberal economists, tell us quickly: how much is a ray of sunlight worth? Fresh air without lead or sulphur fumes? A dip in the sea or the lakes?

At what price will industry and the banks be able to ransom all this in order to sell us at retail—in the form of air purifiers, clinics, and hotel rooms—what they have stolen wholesale? And what's the price of hearing, smell, and human life? What, in your cost/benefit analyses, is the benefit that, in spite of everything, will compensate for and make profitable deafness, bladder cancer, and the direct or indirect, whole or partial, extermination of a Third World population? Because if everything has a price, in the end everything can not only be sold, but can also be bought.

All our troubles, said Ivan Illich, come from a market based production system that has no interest in what is *good for everyone*; it only knows exchange values, which are in essence relative. René Dumont says almost the same thing using other examples. The imperialist world as a whole—that composite of countries and classes—has spawned a way of life that can never be attained by the whole of the planet. If the whole world wanted to eat the way North Americans do, we would have to quadruple the agricultural production of the globe. By the end of the century we would have to multiply it by eight. Furthermore, to feed themselves North Americans and Europeans together use around 20% more of the world's agricultural land on top of their own. We are, says Dumont, "murderers who take protein from the mouths of poor children."

This is not a rhetorical flourish. Judge for yourself: of 70 million tons of fish caught in the seas, the Third World consumes 14 million tons, while 25 million are made into meal that ends up "in the troughs of our domestic animals." With every kilo of eggs, chicken, steak, we take four to six kilos of less agreeably tasting protein from the children of the Third World, who are stunted by malnutrition.

And the proof that "our" way of life (created for a privileged minority) cannot be made available to all is that it goes into a crisis as soon as newcomers try to adopt it. You must have noticed it: since the Japanese began eating meat, there has been a shortage in the rest of the world (including North America). Since the Soviet government has begun trying to mitigate the disaster of its agricultural policies with imports, the price of cattle feed has climbed dizzily, further driving up the price of meat (and making it more scarce).

The fact is, all of humanity cannot live like the privileged 20% of North Americans and Europeans, whose consumption style is the—unattainable—standard for the rest of the North Americans and Europeans, as well as the world. There are not enough mineral resources, air, water, or land for the whole world to adopt "our" rapacious ways of production and consumption. It hasn't been so long since "western" technocrats were denying what is now becoming obvious. They believed that if we just exported "our" techniques, "our" way of production and of life would become possible. According to them, for example, the introduction of high-yield rice into India was supposed to bring on a "green revolution" which would avoid revolution altogether. Wrong: the introduction of the high-yield rice has already provoked several riots.

Why? Because these varieties of rice need perfect contouring, soil irrigation and drainage, chemical fertilizers, and insecticides. Which is as much as saying that—lacking a social and cultural revolution, or without a tremendous investment of volunteer labor from peasants organized into cooperatives and communes—cultivation of the high-yield rice is only within the reach of rich farmers. According to Dumont, "60% of the Indian population lives in abject poverty, on less than 10¢ a day"; so how will the poor peasants pay for fertilizer and insecticides? How will they level and drain their land unless they hold it in common? In fact, the "green revolution" condemns them to work for the rich farmers. Which is what they did. And they found that because of the abundance of the supply of people, the price of their labor dropped. Hence the riots.

Moral: exporting capitalist techniques engenders or accelerates capitalist concentration. Since these techniques can never be understood and used by everyone—by the poor or landless peasant majority—their adoption further reinforces the arrogance and power of the rich farmers over the village, which means also over the bureaucrats, the politicians, and the police, who all depend on the wealth of the local landowners. Thus wealth does not spread around, and there is no real development.

Development aid? Development toward what? What "aid missions" try to organize the poor peasants and make new practical and theoretical knowledge available to all? That would be meddling in politics. Teaching? Dumont indicted traditional schooling—a machine for reproducing inequality—even before Ivan Illich.[2] Listen to him again:

> As long as contempt for work persists, all efforts toward a less unequal society will remain at the stage of moral postulates and incantation. To do away with this contempt—which has not been fully achieved in the Soviet Union—would require first of all that everyone participate extensively in some manual work...not on the stupefying assembly line of an automated factory, but at the craftman's workbench, where, by using his hands, a person can develop a kind of skill that is as indispensible as that of abstract reasoning. Diversified work, combined so as to alternate in the factory and in the fields becomes a joy, says William Morris. It will do away with the isolation of the manual from the intellectual, the town from the country...When they have worked with the farmers, my students at the University of Ottawa have things to teach me."

Praise for austerity, frugality, the bicycle, and Chinese socialist civilization; condemnation for the automobile and all it implies. I can hear the protests popping up from this side of the world: "When only a few middle class people had cars, it was hurray for the automobile! Now that the people all begin to have them, you condemn driving." That is true.

But that's because the automobile is something the middle class invented for itself. It is worthless unless it's the privilege of the minority. As soon as the majority get one, the antisocial character of the car becomes clear. It loses its use value; it becomes a source of frustration, danger, expense, and inconvenience for everyone (whether they own one or not). It creates noise, stench, poisons, and suffocated cities made uninhabitable at the center, spreading at their outskirts, gnawing away at the countryside—itself already chopped up by superhighways.

The middle class then deserts the dying cities and increasingly gives up the use of the car. They prefer the airplane, the helicopter, and even railway travel. Denied cars for so long, ordinary people hang on to them even tighter, and are only afraid of being denied them a second time. They still don't notice that the advantages of the bourgeois lifestyle are disappearing and that they are turning into disadvantages *by the very fact that ordinary people can obtain them.* How can this be explained to them, wonders Dumont.

This is what puts the "fraction of the working class that is rising into the middle class, now a majority in rich countries," in opposition to "the present day proletarians who are the rural masses, the inhabitants of shantytowns, and other unemployed people of the dependent countries." How, he asks, can our working class be brought to accept "the disciplines that will be imposed one day by the necessary zero growth of their total production? How can solutions be required of them that are more revolutionary than those proposed by the parties we now call revolutionary?"

How? Why the answer is before you. It is the crisis of the capitalist lifestyle, the impoverishment material growth has brought about, the decay of institutions, the repressiveness of social controls, the ideological and social bankruptcy of the market based production system. All this will open the way to the post-capitalist era by making clear this fact: The only way to live better is to produce less, to consume less, to work less, to live differently.

Dumont says to himself: "We are being cornered by socialism because the profit-based economy is taking *all* of us to our downfall." This is beginning to be felt and understood.

Le Sauvage, March 1973

3. The Social Ideology of the Motorcar

The worst thing about cars is that they are like castles or villas by the sea: luxury goods invented for the exclusive pleasure of a very rich minority, and which in conception and nature were never intended for the people. Unlike the vacuum cleaner, the radio, or the bicycle, which retain their use value when everyone has one, the car, like a villa by the sea, is only desirable and useful insofar as the masses don't have one. That is how in both conception and original purpose the car is a luxury good. And the essence of luxury is that it cannot be democratized. If everyone can have luxury, no one gets any advantages from it. On the contrary, everyone diddles, cheats, and frustrates everyone else, and is diddled, cheated, and frustrated in return.

This is pretty much common knowledge in the case of the seaside villas. No politico has yet dared to claim that to democratize the right to vacation would mean a villa with private beach for every family. Everyone understands that if each of 13 or 14 million families were to use only 10 meters of the coast, it would take 140,000 km of beach in order for all of them to have their share! To give everyone his or her share would be to cut up the beaches in such little strips—or to squeeze the villas so tightly together—that their use value would be nil and their advantage over a hotel complex would disappear. In short, democratization of access to the beaches point to only one solution—the collectivist one. And this solution is necessarily at war with the luxury of the private beach, which is a privilege that a small minority takes as their right at the expense of all.

Now, why is it that what is perfectly obvious in the case of the beaches is not generally acknowledged to be the case for transportation? Like the beach house, doesn't a car occupy scarce space? Doesn't it deprive the others who use the roads (pedestrians, cyclists, streetcar and bus drivers)? Doesn't it lose its use value when everyone uses his or her own? And yet there are plenty of politicians who insist that every family has the right to at least one car and that it's up to the "government" to make it possible for everyone to park conveniently, drive easily in the city, and go on holiday *at the same time as everyone else*, going 70 mph on the roads to vacation spots.

The monstrousness of this demogogic nonsense is immediately apparent, and yet even the left doesn't disdain resorting to it. Why is the car treated like a sacred cow? Why, unlike other "privative" goods, isn't it recognized as an antisocial luxury? The answer should be sought in the following two aspects of driving: 1) Mass motoring effects an absolute triumph of bourgeois ideology on the level of daily life. It gives and supports in everyone the illusion that each individual can seek his or her own benefit at the expense of everyone else. Take the cruel and aggressive selfishness of the driver who at any moment is figuratively killing the "others," who appear merely as physical obstacles to his or her own speed. This aggressive and competitive selfishness marks the arrival of universally bourgeois behavior, and has come into being since driving has become commonplace. ("You'll never have socialism with that kind of people," an East German friend told me, upset by the spectacle of Paris traffic.) 2) The automobile is the paradoxical example of a luxury object that has been devalued by its own spread. But this practical devaluation has not yet been followed by an ideological devaluation. The myth of the pleasure and benefit of the car persists, though if mass transportation were widespread its superiority would be striking. The persistence of this myth is easily explained. The spread of the private car has displaced mass transportation and altered city planning and housing in such a way that it transfers to the car functions which its own spread has made necessary. An ideological ("cultural") revolution would be needed to break this circle. Obviously this is not to be expected from the ruling class (either right or left).

Let us look more closely now at these two points.

When the car was invented, it was to provide a few of the very rich with a completely unprecedented privilege: that of traveling much faster than everyone else. No one up to then had ever dreamt of it. The speed of all coaches was essentially the same, whether you were rich or poor. The carriages of the rich didn't go any faster than the carts of the peasants, and trains carried everyone at the same speed (they didn't begin to have different speeds until they began to compete with the automobile and the airplane). Thus, until the turn of the century, the elite did not travel at a different speed from the people. The motorcar was going to change all that. For the first time class differences were to be extended to speed and to the means of transportation.

This means of transportation at first seemed unattainable to the masses—it was so different from ordinary means. There was no comparison between the motorcar and the others: the cart, the train, the bicycle, or the horse-car. Exceptional beings went out in self-propelled vehicles that weighed at least a ton and whose extremely complicated mechanical organs were as mysterious as they were hidden from view. For one important aspect of the automobile myth is that for the first time people were riding in private vehicles whose operating mechanisms were completely unknown to them and whose maintenance and feeding they had to entrust to specialists. Here is the paradox of the automobile: it appears to confer on its owners limitless freedom, allowing them to travel when and where they choose at a speed equal to or greater than that of the train. But actually, this seeming independence has for its underside a radical dependency. Unlike the horse rider, the wagon driver, or the cyclist, the motorist was going to depend for the fuel supply, as well as for the smallest kind of repair, on dealers and specialists in engines, lubrication, and ignition, and on the interchangeability of parts. Unlike all previous owners of a means of locomotion, the motorist's relationship to his or her vehicle was to be that of user and consumer—and not owner and master. This vehicle, in other words, would oblige the "owner" to consume and use a host of commercial services and industrial products that could only be provided by some third party. The apparent independence of the automobile owner was only concealing the actual radical dependency.

The oil magnates were the first to perceive the prize that

could be extracted from the wide distribution of the motorcar. If people could be induced to travel in cars, they could be sold the fuel necessary to move them. For the first time in history, people would become dependent for their locomotion on a commercial source of energy. There would be as many customers for the oil industry as there were motorists—and since there would be as many motorists as there were families, the entire population would become the oil merchants' customers. The dream of every capitalist was about to come true. Everyone was going to depend for their daily needs on a commodity that a single industry held as a monopoly.

All that was left was to get the population to drive cars. Little persuasion would be needed. It would be enough to get the price of a car down by using mass production and the assembly line. People would fall all over themselves to buy it. They fell over themselves all right, without noticing they were being led by the nose. What, in fact, did the automobile industry offer them? Just this: "From now on, like the nobility and the bourgeoisie, you too will have the privilege of driving faster than everybody else. In a motorcar society the privilege of the elite is made available to you."

People rushed to buy cars until, as the working class began to buy them as well, defrauded motorists realized they had been had. They had been promised a bourgeois privilege, they had gone into debt to acquire it, and now they saw that everyone else could also get one. What good is a privilege if everyone can have it? It's a fool's game. Worse, it pits everyone against everyone else. General paralysis is brought on by a general clash. For when everyone claims the right to drive at the privileged speed of the bourgeoisie, everything comes to a halt, and the speed of city traffic plummets—in Boston as in Paris, Rome, or London—to below that of the horsecar; at rush hours the average speed on the open road falls below the speed of a bicyclist.

Nothing helps. All the solutions have been tried. They all end up making things worse. No matter if they increase the number of city expressways, beltways, elevated crossways, 16-lane highways, and toll roads, the result is always the same. The more roads there are in service, the more cars clog them, and city traffic becomes more paralyzingly congested. As long as there are cities, the problem will remain unsolved. No matter how wide and fast a superhighway is, the speed at which vehicles can come off it to enter the city cannot be greater than the average speed on the city streets. As long as the average speed in Paris is 10 to 20 kmh, depending on the time of day, no one will be able to get off the beltways and autoroutes around and into the capital at more than 10 to 20 kmh.

The same is true for all cities. It is impossible to drive at more than an average of 20 kmh in the tangled network of streets, avenues, and boulevards that characterize the traditional cities. The introduction of faster vehicles inevitably disrupts city traffic, causing bottlenecks—and finally complete paralysis.

If the car is to prevail, there's still one solution: get rid of the cities. That is, string them out for hundreds of miles along enormous roads, making them into highway suburbs. That's what's been done in the United States. Ivan Illich sums up the effect in these startling figures: "The typical American devotes more than 1500 hours a year (which is 30 hours a week, or 4 hours a day, including Sundays) to his [or her] car. This includes the time spent behind the wheel, both in motion and stopped, the hours of work to pay for it and to pay for gas, tires, tolls, insurance, tickets, and taxes...Thus it takes this American 1500 hours to go 6000 miles (in the course of a year). Three and a half miles take him [or her] one hour. In countries that do not have a transportation industry, people travel at exactly this speed on foot, with the added advantage that they can go wherever they want and aren't restricted to asphalt roads."[1]

It is true, Illich points out, that in non-industrialized countries travel uses only 3 to 8% of people's free time (which comes to about two to six hours a week). Thus a person on foot covers as many miles in an hour devoted to travel as a person in a car, but devotes 5 to 10 times less time in travel. Moral: The more widespread fast vehicles are within a society, the more time— beyond a certain point—people will spend and lose on travel. It's a mathematical fact.

The reason? We've just seen it: The cities and towns have been broken up into endless highway suburbs, for that was the only way to avoid traffic congestion in residential centers. But the underside of this solution is obvious: ultimately people can't get around conveniently because they are far away from everything. To make room for the cars, distances have increased. People live far from their work, far from school, far from the supermarket—which then requires a second car so the shopping can be done and the children driven to school. Outings? Out of the question. Friends? There are the neighbors...and that's it. In the final analysis, the car wastes more time than it saves and creates more distance than it overcomes. Of course, you can get yourself to work doing 60 mph, but that's because you live 30 miles from your job and are willing to give half an hour to the last 6 miles. To sum it all up: "A good part of each day's work goes to pay for the travel necessary to get to work." (Ivan Illich)

Maybe you are saying, "But at least in this way you can escape the hell of the city once the workday is over." There we are, now we know: "the city," the great city which for generations was considered a marvel, the only place worth living, is now considered to be a "hell." Everyone wants to escape from it, to live in the country. Why this reversal? For only one reason. The car has made the big city uninhabitable. It has made it stinking, noisy, suffocating, dusty, so congested that nobody wants to go out in the evening anymore. Thus, since cars have killed the city, we need faster cars to escape on superhighways to suburbs that are even farther away. What an impeccable circular argument: give us more cars so that we can escape the destruction caused by cars.

From being a luxury item and a sign of privilege, the car has thus become a vital necessity. You have to have one so as to escape from the urban hell of the cars. Capitalist industry has thus won the game: the superfluous has become necessary. There's no longer any need to persuade people that they want a car; it's necessity is a fact of life. It is true that one may have one's doubts when watching the motorized escape along the exodus roads. Between 8 and 9:30 a.m., between 5:30 and 7 p.m., and on weekends for five and six hours the escape routes stretch out into bumber-to-bumper processions going (at best) the speed of a bicyclist and in a dense cloud of gasoline fumes. What remains of the car's advantages? What is left when, inevitably, the top speed on the roads is limited to exactly the speed of the slowest car?

Fair enough. After killing the city, the car is killing the car. Having promised everyone they would be able to go faster, the automobile industry ends up with the unrelentingly predictable result that everyone has to go as slowly as the very slowest, at a speed determined by the simple laws of fluid dynamics. Worse: having been invented to allow its owner to go where he or she wishes, at the time and speed he or she wishes, the car becomes, of all vehicles, the most slavish, risky, undependable and uncomfortable. Even if you leave yourself an extravagant amount of time, you never know when the bottlenecks will let you get there. You are bound to the road as inexorably as the train to its rails. No more than the railway traveler can you stop on impulse, and like the train you must go at a speed decided by someone else. Summing up, the car has none of the advantages of the train and all of its disadvantages, plus some of its own: vibration, cramped space, the danger of accidents, the effort necessary to drive it.

And yet, you may say, people don't take the train. Of course! How could they? Have you ever tried to go from Boston to New York by train? Or from Ivry to Treport? Or from Garches to Fountainebleau? Or Colombes to l'Isle-Adam? Have you tried on a summer Saturday or Sunday? Well, then, try it and good luck to you! You'll observe that automobile capitalism has thought of everything. Just when the car is killing the car, it arranges for the alternatives to disappear, thus making the car compulsory. So first the capitalist state allowed the rail connections between the cities and the surrounding countryside to fall to pieces, and then it did away with them. The only ones that have been spared are the high-speed intercity connections that compete with the airlines for a bourgeois clientele. There's progress for you!

The truth is, no one really has any choice. You aren't free to have a car or not because the suburban world is designed to be a function of the car—and, more and more, so is the city world. That is why the ideal revolutionary solution, which is to do away with the car in favor of the bicycle, the streetcar, the bus, and the driverless taxi, is not even applicable any longer in the big commuter cities like Los Angeles, Detroit, Houston, Trappes, or even Brussels, which are built by and for the automobile. These splintered cities are strung out along empty streets lined with identical developments; and their urban landscape (a desert) says, "These streets are made for driving as quickly as possible from work to home and vice versa. You go through here, you don't live here. At the end of the workday everyone ought to stay at home, and anyone found on the street after nightfall should be considered suspect of plotting evil." In some American cities the act of strolling in the streets at night is grounds for suspicion of a crime.

So, the jig is up? No, but the alternative to the car will have to be comprehensive. For in order for people to be able to give up their cars, it won't be enough to offer them more comfortable mass transportation. *They will have to be able to do without transportation altogether* because they'll feel at home in their neighborhoods, their community, their human-sized cities, and *they will take pleasure in walking from work to home*—on foot, or if need be by bicycle. No means of fast transportation and escape will ever compensate for the vexation of living in an uninhabitable city in which no one feels at home or the irritation of only going into the city to work or, on the other hand, to be alone and sleep.

"People," writes Illich, "will break the chains of overpowering transportation when they come once again to love as their own territory their own particular beat, and to dread getting too far away from it." But in order to love "one's territory" it must first of all be made *livable*, and not *trafficable*. The neighborhood or community must once again become a microcosm shaped by and for all human activities, where people can work, live, relax, learn, communicate, and knock about, and which they manage together as the place of their life in common. When someone asked him how people would spend their time after the revolution, when capitalist wastefulness had been done away with, Marcuse answered, "We will tear down the big cities and build new ones. That will keep us busy for a while."

These new cities might be federations of communities (or neighborhoods) surrounded by green belts whose citizens—and especially the schoolchildren—will spend several hours a week growing the fresh produce they need. To get around everyday they would be able to use all kinds of transportation adapted to a medium-sized town: municipal bicycles, trolleys or trolley-buses, electric taxis without drivers. For longer trips into the country, as well as for guests, a pool of communal automobiles would be available in neighborhood garages. The car would no longer be a necessity. Everything will have changed: the world, life, people. And this will not have come about all by itself.

Meanwhile, what is to be done to get there? Above all, never make transportation an issue by itself. Always connect it to the problem of the city, of the social division of labor, and to the way this compartmentalizes the many dimensions of life. One place for work, another for "living," a third for shopping, a fourth for learning, a fifth for entertainment. The way our space is arranged carries on the disintegration of people that begins with the division of labor in the factory. It cuts a person into slices, it cuts our time, our life, into separate slices so that in each one you are a passive consumer at the mercy of the merchants, so that it never occurs to you that work, culture, communication, pleasure, satisfaction of needs, and personal life can and should be one and the same thing: a unified life, sustained by the social fabric of the community.

Le Sauvage, September-October 1973

4. Socialism or Ecofascism

When the Mansholt memorandum and the Meadows report to the Club of Rome first came out, many of us first reacted with delight. At last capitalism was admitting its crimes. It was admitting that the logic of profit had led it to produce for the sake of production, to demand growth for the sake of growth, to waste irreplaceable resources, to plunder the planet. The logic of profit had made it more and more complicated and costly to satisfy basic needs (to breathe, to get well, to be clean, to have a roof over one's head, to get around, etc.); it had increased people's general frustration as it increased the mass of commodities intended to replace things that were formerly free: air, sunlight, space, forests, the sea....It was admitting that things couldn't go on this way because if they did it would bring on a catastrophe that would threaten the existence of the higher forms of life on the earth. It was recognizing that all the values of capitalist civilization needed to be reexamined. The way we live, consume, and produce had to be changed.

This is the meaning to be found in the Mansholt memorandum and the Meadows report. They brought grist to the mill of all who reject capitalism because of its logic, premises, and consequences. Still, nothing has yet been accomplished. There will be no miracle. Capitalism will not change itself into its opposite because a few very big bosses were touched by grace and recognized the physical limits of growth. On the contrary, if capitalism today admits that there are limits, that the next 30 years cannot be like the last 30, and that the six billion inhabitants of the year 2000 could not possibly adopt our pattern of industrialization—if intelligent capitalism recognizes all this, we may be sure that it is not in order to prepare for its own suicide. Rather, it is in order to prepare itself to fight on new fronts, with new weapons and new economic goals.

What goals? The same ones that the left, which capitalism is now trying to overtake, could have put forward in a revolutionary program of awesome simplicity. While consuming and working *less* we can live *better,* though differently. This statement is easy to prove. We will return to it later. The only question to ask is, *can we live better while consuming less within a capitalist framework?*

Don't be in too much of a hurry to answer or especially to prove (which is theoretically possible) that the answer must be negative. For an organized and self-conscious capitalism will never agree to put the question in this guise. As far as it is concerned, this question must be swept away and an imperative substituted: "We have to get there." From the moment it becomes clear that the pursuit of material growth leads to worldwide dilemmas—and that is undeniable no matter how you quibble over the figures and the time frame—the problem put to capitalism is essentially a practical one. It must either disappear or change the basis and nature of its economic growth.

Will capitalism succeed? It's too soon to tell. But what is already certain is that it is working out theoretical and practical tools that will make it able to face, by means of a great shift, the historical novelty of a real problem. Don't underestimate its capacity for adaptation and its cunning. Don't confuse capitalism with the narrow obstinancy of most owners and managers. It is not they who work out the long term strategies of capital. That is conceived and discreetly put into practice by a few dozen industrial and banking giants, who are so large they are obliged to have a vision. Like everything, they can buy it: they simply order it from the universities, foundations, and research centers.

This is all the Club of Rome did. This select group of international bosses sent the order to the Massachusetts Institute of Technology (MIT). MIT delivered the goods in the form of well-founded recommendations. It's up to economists now to figure out how capitalism can accomodate these recommendations. Let's consider the salient points:

• Starting in 1975, industrial production in "rich" countries must stop growing. Only the industries in "poor" countries should continue to develop, for another 15 years.

• Around 1990, worldwide industrial production will have to have tripled, but consumption of mineral resources will have to be not more than a quarter of current consumption. The following two series of measures can make this happen:

1) A search for maximum product durability. It must become practically impossible for things to wear out, and at the least they must be easy to repair—an end to continual changes of fashion and style, to gadgets and junk.

2) A systematic recovery and recycling of all raw materials, which, like energy, will be allocated according to a strict central plan. Only the production of intangible "goods" will be able to develop freely.

All this seems to be simply good sense. After all, six times less industrial production can get us the same amount of use values we have now if only we learn to make things six times more durable. The distribution of material goods will be just about egalitarian, since they will last more than a generation. We will work less, buy less, and still will not have to deprive ourselves of anything. Who would miss "novelties" if they weren't put on the market? Did you miss color TV before the electronic giants introduced it? Does it enrich you life? Would you miss shoddy, brightly colored men's underwear? And the electric-machine that saves you the trouble-of-exercising-because-it-makes-your-muscles-work-without-your-lifting-a-finger ("You can knit while it works for you"), is that an enrichment, an impoverishment, a deterioration, or what?

The case is made. "Consume less and you'll live more." But if things are that simple, why didn't the capitalists think of this sooner? Why did they first create "affluent" civilization—which in fact is a civilization of poverty in waste—rather than concerning themselves at the outset with "real riches"? And why do they suddenly claim to be concerned with it?

The answer to these questions is implicit in two propositions: 1) to avoid economic crises, advanced capitalism must have waste; 2) advanced capitalism is henceforth obliged to stop certain wastes if it wants to avoid another order of crises, which will be ecological at first, then economic and political later.

Let's examine these two propositions more closely. This will help us appreciate the awesome problems of conversion that industrial nongrowth will impose on capitalism.

A capitalist is not primarily someone who has a fortune and who lives off the work of others. That also describes the slave-owner, the usurer or moneylender, the feudal lord. The essence of a capitalist is that for him money is not primarily something that you spend (spent money by definition is not capital) but something that you invest in order to make a profit which in turn will be invested in order to make an even larger profit and so on forever. The growth of profit, of production, of the company, is the only criterion of success for its managers. And it matters little whether they are owners or salaried executives, bosses by divine right or managerial technocrats. In any case they must act like capitalists—that is, deliver the obsessive, obstinate, tyrannical, message of capital that can't say anything other than "more, bigger, faster."

And why always more, faster? It's quite simple. If you don't invent or buy the new machines with which fewer workers can manufacture more commodities, you can be sure a competitor will put in new machines before you do and will ruthlessly eat away at your share of the market. So you have to get ahead of him. Your profits must always be *at least* as large as your competitors' so that you can always pay off and replace your machines *at least* as quickly as they do.

A different policy—that would favor using the same machines for a long time—assumes the prior elimination of all competition. And that can only be achieved in two ways: production planning by private cartel agreements to which every firm adheres under pain of punishment as terrible as the mafia imposes on an unruly gang, or public planning and social management of all industry.

The increasingly rapid replacement of fixed capital (five years on the average) is part of the capitalist logic of "healthy competition." And this speeding up of innovation becomes even more pronounced when wages tend to go up under pressure from the workers. To avoid the increase in costs, which would lower its profits, capitalism has only one way out—investments in productivity by continual "modernization" of technology, machines, and methods. More, bigger, faster.

But soon another problem surfaces. Who will consume the rising flood of goods which pours out of these more and more efficient factories? How long can this race go on, with everyone trying to get ahead of the others and trying to escape the natural tendency of the profit rate to fall, by stepping up the pace of innovation? Won't growth have to stop because the market will be physically incapable of absorbing additional goods? What a catastrophe that would be for capital. Consumer goods industries would stop growing and investing; capital goods industries would correspondingly slacken. Unemployment would spread and the economy would spiral downward into crisis.

How can this be avoided? It's very simple. To insure that your future products won't be left on your hands, make sure of the speedy deterioration of your past and present products. In other words, arrange things so that people are constantly exchanging the old for the new, either because the worn-out object is not repairable (physical obsolescence) or because big advertising campaigns boast of the superiority of new models and put the old models "out of style" and make them a sign of poverty (moral obsolescence). To be even more sure, most big firms make certain that physical deterioration prevents those people who are disobedient to fashion from keeping the same object too long. The following story is particularly edifying in this regard. The first fluorescent lights put out in 1938 by Philips (Holland) had a lifetime of 10,000 hours. They could "burn" continuously for 14 months. Bad business, decided the Philips management, who, before putting the tubes on the market, carefully reduced the lifetime to 1000 hours (42 days). *The Waste-makers* by Vance Packard contains a number of similar anecdotes.

Again, take this telling example: Suppose that, at a cost of $25 (in leather, work, machine hours), a manufacturer could produce either five pairs of shoes that would each last 300 hours or two pairs of shoes that would last 3000 hours. In the first case, for $25, 1500 hours of wear are created; in the second case, 6000 hours of wear. Which will he choose? The first, obviously. Because first of all on each cheap pair the profit is proportionally higher than on a durable pair. Then, and above all, because the cheap shoes will wear out ten times faster and he can then sell ten times more a year. His profit in the final tally will be 15 times higher than if he made durable shoes.

It matters little to him that he is wasting leather, work, energy, and machines. Maximum profit is not made by economizing the factors of production, but by means of waste and deterioration that guarantee an appropriate capital turnover. With the profits he obtains, the manufacturer has only to invent new styles and new methods of production that will further increase the consumption of shoes.

Repair nothing—use it up and throw it away. Change for change's sake. To give you a taste for it, here are first disposable packages, then disposable fabrics, soon disposable crockery. The beauty of affluence! Prosperity rests on the faster and faster transformation of mountains of junk into mountains of debris. And the happy agents of this transformation, known as consumers, are the same ones who joylessly give their energy to make these things they hope to find time to use between coming home and going to sleep. Isn't this the secret of indefinite growth for capitalism?

Well, no. For the last ten years, more or less, one of the implicit postulates on which capitalism rests is no longer valid. It is no longer true that the more that is produced the lower is the cost of each unit and the larger the nation's wealth. Beyond a certain point it is rather the reverse. Growth destroys more wealth than it creates and the costs, direct or indirect, are going up. All the "overdeveloped" countries have already experienced this. *"The quality of life" goes down even though production increases.* In all industrial areas the physical limit of growth has been reached and the profitability of investments can only decline further. New York, Detroit, Tokyo, the Ruhr valley, and soon Paris are choking on their own congestion. Rivers and lakes have become brown, pestilential muck; chemical smoke poisons the air and damages the respiratory passages. Noise, dirt, and crowding drive out the well-to-do, and the taxes paid by those who are left are not enough to permit the cities to reverse the decline.

In order for production to increase in these areas, the air and water would first have to be purified, at great expense. The environment cannot absorb the waste from any more industries—even if they are supposed to be "clean"—if the rate of pollution from the existing industries is not decreased. Thus the cost of future plants and production will be higher than in the past. Big business will find that it is in the same situation as the car manufacturers who, if they want to continue selling their cars, will have to widen the roads, build new ones, and tear down and remodel the inner cities themselves in order to accommodate their product.

"Let the polluters pay," people say. Of course. But what's the result of that? Higher prices and lower profits. "The bosses are able to pay," they continue. That's certainly true. But then they will make everyone pay. For if capitalists have to invest in "clean technology," one of two things will happen:

• They will finance these investments out of their profits without raising the retail price, in which case their profits go down. This slows down the growth of production or even brings it to a standstill, unemployment spreads, and real wages decline (that's what happened in the United States in the early seventies).

• Capitalists raise their prices in order to keep their profits up. But in this case, since goods become more and more expensive, people will buy relatively less of them. Again, the production of goods will be slowed down in order to carry on the battle against environmental damage.

Thus the result in either case is the same: growth cannot continue at the same pace and in the same way as before. Concern for the "quality of life" is not compatible with the growth of production that has prevailed up to now. Big business knows this very well. Conglomerates, multinational corporations, and the merchant banks have drawn the unavoidable conclusion: the quality of life has to become a profitable business. Instead of desperately hanging on to material production, they must turn more and more to the production of nonmaterial goods. There's no limit to such growth; this is where the future lies.

The Club of Rome, Sicco Mansholt, Robert Lattes say it so candidly that one cannot help but wonder about their second thoughts. But why should they have second thoughts? They are simply realists. The dreamers are all those classical industrialists who claim to be advocates of continued growth even though the price of energy and primary metals is expected to increase tenfold, even though the water shortage requires distillation of the seas or water recycling, even though removing the heat and wastes produced by thermal power stations presents problems which no one can yet answer, even though the necessity for husbanding or even reproducing the environment will weigh more and more heavily on the costs of production.

Even if the figures in the Meadows report are unreliable, the fundamental truth of its thesis remains unchanged. Physical growth has physical limits, and any attempt to push them back (by recycling and purification) only pushes the problem around. For to renew air, water, and metals requires extensive amounts of the very scarcest resource of all—energy—and all the forms of energy that are available to industry involve chemical, thermal, and/or radioactive pollution. In the foreseeable future, energy will be increasingly scarce and expensive.

So the problem is clear. What is needed is a fundamental change in growth itself, so that growth priorities can focus on intangible goods. But what does that mean in concrete terms? And even before that, how will capitalism deal with this change so that it takes place without a profound crisis?

The answer is right in front of you. Just see how the industrialized world is sloughing off its industries and their pollutants to poor countries and continents. Growth in the U.S. automobile industry is hardly occurring at all, except in Spain and Brazil. Fiat is hardly expanding anywhere anymore except in the USSR, Spain, and Argentina. Renault obtains an increasing proportion of its parts from Yugoslavian and Rumanian licensees. Scandinavian furniture is manufactured in Poland; a good proportion of German cameras comes from Singapore; big German chemical firms are establishing new factories in Brazil (again). In ten years São Paolo will be an agglomeration of 20 million inhabitants. A report done by experts of the Rand Corporation says that before the century is out the United States will have all its manufactured products made abroad and will have nothing on their own territory but scientific and service industries. Maybe you are wondering how they will pay for their manufactured products. Why, with their profits, of course—profits that U.S. factories abroad will be bringing in (and already bring in) from all over the world. As the Rand Corporation sees it, Americans will become a nation of bankers, busy mostly at recirculating their profits levied on the work of others. Seen from this perspective, it is easier to make sense of the way the U.S. government is managing the current monetary crisis, and how other countries are reacting to it. For the Germans, Japanese, British, French, and Dutch have the same ambitions as the Americans, even though on a smaller scale. They too want to live off the rest of the world, in the protective shadow of the United States and in competition with it (the one doesn't exclude the other).

What a marvelous scheme! For us, clean air and water, production of non-material goods, leisure, affluence; to Third World countries, if they are well behaved, material production, dirt, pollution, danger, sweat, and exhaustion, along with congested and polluted cities. When the Meadows report looks forward to tripling worldwide industrial production, while recommending zero growth in industrialized countries, doesn't it imply this neo-imperialist vision of the future? And what about us? Are we going to buy this bill of goods? On the pretext of saving (assuming it were still possible) our environment (or what's left of it) are we going to ally ourselves with the inter-

national bosses of the Club of Rome? Are we going to participate in a scheme that allows them, with the aid of defoliants and
napalm when necessary, to poison the Congo and Zambezi
rivers, to ravish the Amazon, to pump Iran dry, and use India's
unemployed to do the jobs that "developed" people refuse?
Bon appetit.

In any event, this exporting of industries and their pollutants
can only be a transitional stage in the preparation for a particular kind of non-growth. It will help multinational firms to
spread the risks, to gain time, to offset the decapitalization of
their domestic industries, and above all *to create the necessary
conditions for a general cartelization.* When the industries of the
whole world are controlled by a small number of firms (300, it is
predicted), these firms will be able to arrange things among
themselves, to divide up the markets, share the mineral resources, fix their prices in common, plan their total production,
use the same technology, and refrain from all competition.

We have seen all this before, during the great depression of
the 1930s. Capitalism can accept non-growth as long as competition is eliminated in favor of a general cartelization that freezes
the power relations among firms, guarantees them their profits,
and substitutes capitalist planning for the market. But let us try
to see further. What can big business do with these guaranteed
profits? Not to invest them would mean that capitalism was
dying, that it had become parasitic, like the mafia. The leaders
of the Club of Rome still think they can do better. Since there
will be no opportunities for profitable new investments in the
production of material goods, why shouldn't they try to take
over and industrialize the production of non-material goods,
most of which are still small scale, precapitalist operations?
Imagine! If medicine, sex, education, and culture were industrialized what a huge field would be opened up for capitalist
growth.

These are not at all crazy ideas. Research is now going
forward at a quick pace on the industrialization of sex. (We will
return to this later.) These ideas are no crazier than the idea of
industrializing the sun, the fresh air, or the scenery would have
been even twenty years ago. But this industrialization is already
reaching its limit. Conglomerates and banks are in the process

of buying up the last sites where you can still enjoy the sun, the sea, and a view for free. And on them they are building airports, apartment towers, hotels with swimming pools, decked out beaches, marinas, and parking lots. So, if you want to stretch out in the sun, *you have got to use (and pay for) these industrial* facilities. Enjoyment of the sun, the beach, and relaxation is made dependent on renting them.

Capitalism has accomplished the feat of *capitalizing* picturesque spots and scenery—that is, it has transformed them into capital—of managing, operating and renting them out to "users." To do this, all they had to do was *industrialize the means of access and use of these sites.* Why not do the same for other "intangible" consumables?

Take medicine. It is still to a great extent a kind of luxury cottage industry. Capital has already persuaded people that they can't take care of themselves or even stay healthy without industrial devices—the majority of which are either placebos or toxic—which they must buy at a pharmacy in complicated packaging and under complicated names. They have also been convinced that there has to be an industrial center for health care, called a hospital, to take care of them (if not to cure them). But, by some shameful mistake, most of those in charge of pharmaceutical products and industrialized health care are still relatively independent of capital.

This outdated state of affairs cannot go on. Doctors and psychiatrists must become employees of capital; their functions must become industrialized. And, undoubtedly, before long we will learn that a conglomerate which controls pharmaceutical laboratories, clinics, manufacturers of electronic medical equipment, and insurance companies has introduced health insurance...with the blessing and financial support of the government. Everyone who subscribes to a "health package," comprising periodic automatic medical examinations, vaccinations, preventive medications, and diets, will be covered against the risks of various illnesses. All the "health products" and equipment will of course be manufactured by the conglomerate and prescribed by hired doctors for whose education the conglomerate will have paid.

While we're doing health, why not industrialize sex? Professor John Postgate of the University of Sussex sets out some fairly detailed ideas about this in the *New Scientist* of April 1973. To reduce population growth, Postgate proposes a pill that would allow couples to have only boys. Given the dominant phallocracy, Postgate thinks most couples would choose to have boys only, the result being that the world would end up with five or fifty times as many men as women. Automatic consequence: sharp reduction of the birth rate. Concomitant result: homosexuality and especially masturbation would become prevalent. Since Postgate has the industrial spirit, he doesn't say "nothing will be left for men except to masturbate"; he says "mechanical and pictorial substitutes for normal sex practices could be widely used." And there you have the sex industry. The mechanical and pictorial substitutes will be quickly perfected; electric, electronic (as we shall see they already exist) and chemical devices will appear; vending machines for masturbating will embellish the corridors of porno movie houses (whose notable achievement has been to save the motion picture industry from the crisis precipitated by television).

As you can see, the idea is always the same. People have to be kept from satisfying their needs in a spontaneous and independent way. They must depend for their satisfaction on institutional and industrial objects that they can only get by *buying* or *renting* from institutions that control them in what Illich calls a "radical monopoly."

Why not go all the way down this primrose path? Why doesn't capital also take over control of prostitution and industrialize it instead of leaving it to the craft, the mafia, and the police? To do this, all that would be needed would be to give the profession its own certification. In a society that has already codified and professionalized all know-how and that has already given academic institutions a radical monopoly on the transmission of skills (at least the socially recognized ones), all they would have to do would be to authorize in the same way the creation of a Bachelor of Sexual Skill (BSS). The prostitution industry would be born simultaneously with the new professional competence which, certified by an academic title, would be a valuable source of new inequalities. There would be those with the BSS and those without, which would make it easier to differentiate among a population that is mostly idle and living on public assistance.[1] Its organization could be based on the domination of supermales. It wouldn't be the first time.

Wouldn't all this follow the logic of schooling? Isn't its function to break spontaneous reactions? To insert between the desire to learn and the possibility of learning a heavy institutional apparatus that is both selective and disciplinary, and which schools rather than teaches or educates? School is the essential machinery for reproducing the social order. Why shouldn't it just go ahead and start teaching the very littlest ones to walk and talk? Think of the market that would open up for the more or less nontangible goods industries: audio-visual equipment to teach speech, and transistorized electro-mechanical equipment to teach walking would be added to the splendid teaching machines which, at last, truly allow the industrialization of books, teaching, and "Kulture."

Will you say these are crazy ideas? Watch out: ideas like these are being spread by an influential group of psychiatrists who consider people who revolt against this increasing "techno-fascism" crazy. Take, for example, Dr. Frank Ervine, a Boston psychiatrist who proposes to lobotomize—that is, to destroy the creative and cognitive faculties by brain surgery—people whose behavior goes beyond "an acceptable level of violence." A hundred or so of these "psychosurgeons" (U.S. and European) are currently going ahead with this kind of cerebral mutilation, particularly with prisoners, "crazy" people, difficult children, and women.[2]

Another example is Dr. Robert Heath of Tulane University who has succeeded in reversing sexual behavior with electrodes implanted in the brain. Implanting some 25 electrodes in a few of his patients, the doctor creates a kind of zombie who is telecommanded by Hertzian waves. In others, the electrodes are connected to transistorized "pleasure packs" which allow the "patients" to approximate orgasm up to 1000 times an hour. Of course, this keeps them completely docile, which is the idea.

But the chief pioneer in "physical control of the mind" is Dr. Jose Delgado, theoretician of a "psycho-civilized society" in which people's behavior, feelings, and actions will be controlled from a distance via a central computer, somewhat like a space ship. He means, in the end, to make people into robots, run by a computer that will assure universal order. Who will program the computer? You guessed it: a committee of psychiatrists, the sole custodians and guarantors of mental health—"We are in the process of creating a civilization wherein those who diverge from the norm are risking brain mutilation," writes Dr. Peter Breggin.[3]

"It is not at all out of the question," writes Ivan Illich, "that, terrorized by the dangers that threaten them, people will deliver themselves over to the technocrats who will take responsibility for keeping growth just this side of the threshold of the destruction of life. This technocratic fascism will also assure maximum subordination of people to tools both as producers and as consumers. People will survive but under conditions that remove all value from life. We will be locked up from cradle to grave in a universal school and a universal hospital, which will be indistinguishable but for their names from a universal prison...The principal task of these engineers will be to manufacture the kind of person who is adaptable to this situation."

We know from the psychosurgeons that this is now physically possible.

What plan can we oppose to these sinister engineers of the soul? That of a society of individuals who, while freely associating for common ends, will have the maximum individual and collective freedom. Obviously this assumes the undermining not only of property, but also of the nature of production technology, the means of production, and the forms of productive cooperation.[4] For it is an illusion to think that notions of "voluntary cooperation," of "democratic planning," of "worker management" will ever be able to have any meaning in a factory where 20,000 workers make tires for an entire country. Such a factory drains the workforce of an entire city or region and this obliges it to depend for everything else on unknown workers, faraway factories, and anonymous bureaucrats.

No, I am not advocating a return to subsistence agriculture or to local self-sufficiency. The point is to reestablish the balance between institutional production and the autonomy of the basic communities. Let's look at the example of the shoes once more from this angle. Assume that institutionalized social production does not involve more than four or five basic models of shoes which are very longlasting and answer the needs people have expressed. So much for the necessities. They can be centrally planned, and their production can be met while reducing the work week of both workers and the shoe factories to 10 or 20 hours.

For everything else—for the non-necessities, for the stylish, for the superfluous shoes—all across the country you would find hundreds of workshops, open day and night, equipped with handy, well-made machines that are easy to repair and use. There you would make shoes to your own taste yourself (after paying for the raw materials). You would be familiar with this from childhood. To make clothing and shoes, to work leather and clay, to shape and fit wood and metal, to make vegetables grow, are all part of basic education, as are electricity and mechanics.

There: that's the whole show. The central plan and its bureaucracy are reduced to very little, which allows for the functioning of a large sector that is *free but not commercial* and thanks to which individuals, neighborhoods, and communities can fashion in their own way their life and their environment, which, at last, are *their own.*

"The general crisis," writes Ivan Illich, "can only be overcome by scaling down the size of tools and power within society."

Le Sauvage, July-August 1973

5. Twelve Billion People?

Despite the increase in the number of trawlers, the annual fish catch has dropped by 11% since 1970. Despite the "green revolution," the per capita production of grains in the Third World has fallen to below the level it was between 1961 and 1965. As a result of the drought in the United States, this year [1974] grain production will fall between 12 and 19%. As a result of the floods in Bangladesh, grain production will fall by at least an eighth. In 1961 the worldwide grain reserves were equivalent to more than three months consumption needs; now they wouldn't last even four weeks.

We aren't headed for famine anymore, we are there. Last year around 70 million people died of malnutrition or hunger. This figure is cited by Swedish Nobel laureate Normann Borlaug, a major promoter of the "green revolution," who fears that in the coming year 10 to 50 million more people may die of hunger in India alone. In the state of Bihar (India) an epidemic of smallpox has just killed 25,000 people.

This is the background of the World Conference on Population, sponsored by the UN, which is being held in Bucarest. Is overpopulation the cause of famine? of underdevelopment? of wars? Aside from a few clumsy theories, no one claims this to be the case. The majority of the Third World and all the socialist countries have energetically maintained the opposite. John Rockefeller himself, a supporter of birth control for forty years, wanted to make it clear that "population expansion doesn't create the problems that beset many countries. It makes them worse and multiplies them."

Was the issue going to be killed that easily? Not quite. Even the Chinese representative stated: "There are problems specific to population. We are not denying the importance of a population policy. China has hers; but it is part of a general development plan for the country."

The reality of these specific population problems is best illustrated by two extreme examples: the Sahel and Bangladesh. The current famine in the Sahel, which has numerous causes (climatic, political, social), would never have grown to such proportions if the pasturelands at the borders of the Sahara had not been overgrazed as a result of the increase in the nomad populations. Once the vegetation was stripped from the land, the Sahara began to nibble away at it, moving south at a speed of 9 to 50 km a year. Retreating as the desert advances, the nomads and their flocks have created devastating pressure on new areas. Now only some united action that goes far beyond starvation relief can prevent the further spread of disaster.

In Bangladesh, similarly, the catastrophe's causes are not simply natural ones. On the contrary, for the past 25 years—as a result not only of population pressure, but also as an effect of the "green revolution"—the foothills of the Himalayas have been subjected to intensive deforestation. The ground can't hold the rains anymore, so the run-off carries the land away, and the Ganges and the Brahmaputra unexpectedly flood because their beds are higher. This is the main cause of the catastrophic floods of the last few years. Here, too, a unified plan is necessary, primarily a plan of reforestation comparable to what China has carried out for the past 25 years. Birth control will not be enough, even though it is indispensible if Bangladesh is not to perish well before reaching the 220 million mark (3 times the current number) that demographers predict for the year 2000.

Given these examples, there is a strong temptation to dodge the issue of the world's population in favor of dealing only with that of the most populated regions and countries. The majority of the delegates from the Third World have given in to this temptation. Of what concern is overpopulation to Gabon, which has three inhabitants per square kilometer? Of what concern is it to Brazil, which hopes to surpass the United States and populate its empty lands? Of what concern is it to Argentina, which hopes to double its population in 25 years so as to withstand the pressure from Brazil? How does it concern the USSR, which, worried by the "yellow peril," wants to increase the population of its Asian republics?

When the problem is broken down in this way, it falls quickly into a classic game theory scenario—"the tragedy of the commons." This is it: to make sure that "others" don't get more out of a common pastureland than one does oneself, everybody strains to put the largest number of cows on it as quickly as possible. Result: the grazing land is ruined and all the cows die. This scenario has been played out over whale hunting, and lately is proving to be relevant to the fishing of anchovy, tuna, cod, herring, etc. There is danger that it will be repeated in other fields. That's why the UN has been trying to convince all governments that they have a common interest in slowing population growth.

For if this continues at its current rate, there will be 9 billion people in 1995, 40 billion in 2025, and 100 billion in 2075. The catastrophe will occur well before then—at the beginning of the next century.

If, instead of continuing at its current exponential pace, population growth stabilizes at its present rate of 2% per year, there will be 6.5 billion people on the earth in 1998 (twice as many as in 1965), and 27 billion in 2070. Catastrophe will still be inevitable. The modest objective of the world conference sponsors was to have not more than 12 to 16 billion inhabitants as we approach the year 2100. That would be three or four times the current population.

This seemingly modest objective will actually be very difficult to attain. For, at its current rate of growth, world population will top 12 billion by the year 2035. It is unlikely that the earth can adequately feed a population of that size over time.

For if a world population of double the current size is to have a food allotment that is even half of what Europeans or North Americans have today, you would have to get a European-style, high-intensity yield from all the arable land on earth. To feed a population three times the current size, people would either have to be content with a third of the current European allotment, or else a European yield would have to be obtained from the cultivation of lands that are now still covered with forests. Is this possible? No, at least not for a long time. Agronomists have no trouble pointing out inconsistency when technocrats speak of extending our mechanized, chemicalized agriculture all over the globe.

A few figures will give an idea of the dilemma. It took modern agriculture only 70 years, from 1882 to 1952, to destroy half the topsoil on 38.5% of all cultivated land. During this period the amount of land which could no longer be cultivated increased by 3.45 billion acres. More than a third of the forests that were standing in 1882 have been razed (that is, 4.75 billion acres). Of the 3 billion acres currently under cultivation there are only 1.25 billion acres left of "good land."

Current methods of agriculture are even more destructive than the methods used during that 70 year period. The high per acre yields in North America and Europe are obtained at the price of increasing the expenditure of energy and the disturbance of the water, nitrogen, and carbon cycles, which is untenable in the long run. There are water shortages everywhere. The dilemma is further aggravated by the energy crisis. In 1945, the United States used one calorie of fossil fuel energy to make 3.7

food calories. Today that ratio has fallen to 1:2.8.

The "green revolution" has only been possible in industrialized countries by sharply raising the inputs of fossil fuel, which is limited and irreplaceable. The special new seeds, which were supposed to triple the per acre yield, are actually fragile types that require a synthetic environment in order to thrive. In the United States this environment is created at the expenditure of an energy equivalent to 500 gallons of oil per acre per year.[1]

This explains the failure of the "green revolution" in the Third World. Only the rich farmer could afford the fertilizers, insecticides, machinery, and irrigation that the new seeds require. Hence the acceleration of the flight from the land and the rise in unemployment. By the end of the 1960s the "green revolution" in India had produced a 50% increase in the grain harvest. However, 40% of the increase came from the seeding of new areas, a large part of which had previously been used for growing leguminous plants such as lentils or beans—which are the main source of protein for the Indians. On balance: Indians today do not have more grain per capita than 10 to 15 years ago, but their leguminous allotment has dropped by 30%.

That's not all. After several years in which new wells were drilled and electric water pumps installed, the drop in the level of ground water brought about disastrous (and predictable) droughts in many regions of India. The Philippines, which was banking heavily on the new rice variety IR-8, experienced a different disaster. Because their genetic base is so narrow, the new varieties of grain were subject to the massive spread of diseases and parasites. Tungro (a viral disease), which in 1972 devastated a quarter of the rice fields in the Philippines, wiped out the plans for making this country a heavy grain exporter.[2]

This year there is a new disaster: a shortage of nitrogen fertilizer, which is indispensible to the new varieties of grain. It takes 3 tons of oil to produce one ton of fertilizer. India is no longer in a position to pay for the fertilizer Japan has been selling her, nor even to buy the oil necessary to run even half her fertilizer factories.

The leap forward of agricultural production to beyond its ecological limits is thus colliding with the energy problem. If the whole world were to use U.S. agricultural methods on all the land currently under cultivation, agriculture alone would use up the known oil reserves in a matter of 29 years. The means to feed 8, 12, 16 billion people is still to be found. There is no guarantee that this is possible.

Even so, when the "First World," led by the United States and Sweden, sounds the alarm and calls for population control, the first reaction on the part of the Third World is irritation or rebellion. This oughtn't to be surprising. For with only 13% of the world population, the industrialized capitalist countries consume 87% of the world's energy. They take for themselves half the world's fish, leaving only a fifth to the Third World. To feed themselves they use 20% of the arable land on the globe *beyond what is their own.* They are now setting up in the Sahel, right in the middle of the famine, 375,000 acres of cattle farms meant to provide meat for Europe. They give two-thirds of the world soybean harvest to their animals, even though soy is the main protein food for one billion of the inhabitants of Asia. They use 800 to 900 kilos of grain a year per capita to fatten livestock and poultry, while 150 to 200 kilos would be enough for a Third World resident to put food on the table and feed the chickens. It is said that the hydrosphere and the atmosphere will be poisoned by the wastes of the 8, 12, 16 billion people of the next century; but the 500 million inhabitants of western Europe and North America currently cause the environment as much damage as 10 billion Indians would (if they existed).

Hence the suspicion that when we ask the Third World for population control it may simply be so that we can continue to pillage the planet. For our recommendations to have credibility our societies will have to begin by putting an end to the plunder and stop maintaining or establishing systems that hinder all independent development in the Third World.

Josué de Castro was one of the first to show this.[3] Birth control and sterilization campaigns, as well as the distribution of contraceptives, are effective and sensible (the Indian government knows something about this) only if, along with an overall development policy, they are able to hasten the achievement of a standard of living that encourages a spontaneous drop in the birth rate. In the end that is all John Rockefeller was saying. But what he didn't say, and what others have had to say in his place, is that a development policy begins with agrarian reform. It begins by marshalling the unemployed (20 to 30% of the population) against the causes of "natural" disasters and with campaigns of reforestation, drainage, and soil improvement. It begins with the emancipation of women. All this is what First World intervention, whether military or not, has obstructed for the past twenty years in Guatemala, in the Congo (Zaire), in South Vietnam, in Brazil, in the Dominican Republic, in Indonesia, in the Philippines, in Chile.... As long as the First World continues to subsidize and arm regimes that starve their people and export their "colonial products," its fears of overpopulation will be viewed with suspicion in the Third World. For all that, the fears are well founded.

2 September 1974

REINVENTING THE FUTURE

1. Ezra Mishan, *The Costs of Economic Growth* (New York: Praeger, 1967).

2. In 1936 the first socialist government in the history of France granted all workers two weeks of so-called "legal vacations" with full pay.

3. The weekly magazine of the French Communist Party.

4. Ivan Illich, *Tools for Conviviality* (New York: Harper & Row, 1973), pp. 50-54.

5. Stephen Marglin, "What Do Bosses Do?" in Gorz, ed., *The Division of Labour* (London: Harvester Press, 1977).

6. *Tools for Conviviality, op. cit.*, p. 25.

7. *Ibid.*, p. 34.

AFFLUENCE DOOMS ITSELF

1. René Dumont, *Utopia or Else* (London: Deutsch, 1974). See also Harry Rothman, *Murderous Providence: A Study of Pollution in Industrial Societies* (London: Rupert Hart-Davis, 1972).

2. See *Terres Vivantes*, Plon, 1961, and *L'Afrique Noir Est Mal Partie* (Paris: Le Seuil, 1969).

THE SOCIAL IDEOLOGY OF THE MOTORCAR

1. Ivan Illich, *Energy and Equity* (New York: Harper & Row, 1974).

SOCIALISM OR ECOFASCISM

1. A seventh of the population of New York—more than a million people—lives on public assistance; this figure has nowhere to go but up.

2. Psychosurgery is a more refined and efficient technique than that developed by Dr. Skinner and which in the U.S. and Great Britain is used on prisoners, the violent, and homosexuals. *A Clockwork Orange* described Skinner's method and its effects very accurately.

3. See *Liberation* (New York), vol. 17, no. 7 (October 1972).

4. What Marxists call the relations of production.

TWELVE BILLION PEOPLE?

1. David Pimentel, *Science* (3 November 1973).

2. George Borgstrom, *Focal Points* (New York: Macmillan, 1973).

3. Josué de Castro, *The Geopolitics of Hunger*, revised edition, (New York: Monthly Review Press, 1977).

Chapter III

THE LOGIC OF TOOLS

1. Nuclear Energy: a preeminently political choice

It is estimated that from now to the end of the century 3500 nuclear reactors will be built in the world, at a cost of two trillion dollars. Within 25 years these reactors will be obsolete and new ones, undoubtedly more sophisticated and more expensive, will have to be built. This is an unprecedented and long-term opportunity for profitable investments of unprecedented amounts of capital.

Thanks to nuclear energy, American technology, spurred by the two largest financial groups in the world, will extend its hegemony over the planet. Companies with this technology at their disposal will find themselves a part of the tight net with which the two U.S. multinationals cover the earth. Proud vassals of U.S. corporations, these firms along with their American allies will dominate the Third World countries, where political and technological dependency will assure them maximum income with minimum risk.

The parliament is currently debating a French nuclear program that would fit well into this overall conception of a multinational network. Above all, this program represents a political choice, consistent with the strategy of the biggest corporations of French capitalism: Saint-Gobain-Pont-à-Mousson, Pechiney, Schneider.

Decisions of this importance are not known to be debated in public and put to a vote—it would be too risky. This is why, as Julien Schvartz, UDR (Union Democratique de la Rénovation) deputy from Moselle, remarks, the current debate in parliament is "nothing but a put-on," intended to give a "ridiculous legitimacy" to the "technocratic decisions" that have initiated the present policy.[1]

The enormous political and financial interests at stake account for the intensity of the campaign waged by the "nuclear lobby." This lobby has been particularly untroubled by public scrutiny and criticism because it includes public companies and organizations as well as private ones, and because the multinationals have not had to argue the case for nuclear power themselves (as not long ago they had to argue the case for oil). Nuclear power has been put forward mainly by government officials and bureaucrats who have been happy to hide behind the technical arguments of engineers who are in love with the "big machines."

This is how a fundamentally political choice can be presented as a technical option, an option apparently endorsed by allegedly impartial scientists. The public—that is to say, everyone—is invited to yield to the opinion of the experts and to give them their trust. All objections are discarded as unenlightened, and any inclination toward democratic and popular control is brushed aside on the pretext that the technical complexity of the question makes it the exclusive province of specialists. From the very beginning the nuclear option has been described as incompatible with democracy.

In the end, however, this technocratic arrogance on the pronuclear side has put the best arguments in the hands of their opponents: by hiding behind the "impartiality" of science, the advocates of nuclear energy have aroused the suspicions of many scientists. These scientists have looked more closely and have

discovered with indignation the biases, mistakes, or lies underlying many of the arguments put forth in favor of atomic energy.
Thanks to them, the following problems have come to light:

• Analysis of the accident risk has been largely arbitrary.

• The principal safety mechanism of the reactors has never
been proven reliable.

• Storage of wastes is an unresolved question and industry
has not yet been able to develop the officially chosen method.

• The 85% load factor on which the French program relies
has yet to be reached in any power station. The record to date is
68%.

• Contrary to official information, when nuclear power is
intended for private or industrial heating it will save not 1.5
million tons of fuel oil, but only 500,000 tons for every 1000 Mw.

• The official comparison of net cost fails to take into
account the (particularly heavy) cost of distributing the power.

• There is no assurance of a supply of enriched uranium for
the 1979-1981 period; by 1979 a second factory for isotope
separation will have to be started up (effective cost: three billion
current dollars), which itself will consume the energy produced
by three or four reactors.

• The nuclear program may actually consume more energy
than it produces. The procedure for reckoning the energy balance
sheet of the power stations is the subject of a highly interesting
methodological argument.

The arguments among the experts and disagreements among
scientists have had one major virtue: they have shown the public
that specialists have no absolute authority and that science is a
servant, not a master. It is neutral only in that it can be put to
the service of any cause. Science can shape the means, it cannot
define the goals. Goals depend on the ethical and political
choices of the people themselves.

In short, the disagreements among scientists give the freedom
of choice back to the people and confront them with fundamental political questions:

• What kind of growth does a nuclear program serve?

• Can the quasi-military operation of a nuclear power plant
and the permanent police surveillance of those who work there,
their families, and the surrounding population be compatible

with basic political rights?

• Would we create more or less employment if instead of a nuclear program we invested in renewable energy sources and energy conservation—that is, thermal insulation, greater product durability, product design for repair and recycling, improvement of social services, etc.?

• Would we be better or worse off if economic and cultural development were not based on increased energy consumption?

These questions, and many others, will still be unanswered when the parliamentary debate is over. What is called for is the cooperation of the experts to collect the facts so that the answers can be systematic and consistent. But the answers themselves cannot be supplied by specialists. They involve the choice of the kind of society and civilization we want. The answers are political in the deepest sense of the word.

17 May 1975

2. From Nuclear Electricity to Electric Fascism

The nuclear energy program does not rest on a technological choice; it arises from choices that are political and ideological. Nuclear installations are not one means to ends that might be attained in other ways. They are a means that predetermines which ends are to be reached and that irrevocably prescribes a particular kind of society, to the exclusion of all others.

Beneath its technological outer form the nuclear option has a hidden agenda, which has been worked out by political and business leaders, but of which most of us are unaware. It is this hidden agenda, more than the direct perils of nuclear energy, that we have to understand and fight. It illustrates better than anything one might devise the logic and the direction of French and world capitalism in the current phase of the crisis.

Total Electricity

The decision that France should have a large nuclear industry was taken well before the "oil crisis" of October 1973. It goes back more than 10 years, from the time when EDF (Electricité de France)[1] introduced "total electricity," an entirely electric heating system. From a strictly economic point of view, the prejudice in favor of electrification was dazzlingly senseless. Rather than burning fuel (oil or coal) in domestic and public heating systems, retaining a heat yield of 85%, the idea was to change these fuels into electric current at the power plant, retaining a yield of 30%, then to carry and distribute this current in spite of a high loss rate and a heavy capital investment, and finally to transform the electricity back into heat.

In this way *barely a quarter* of the thermal energy originally consumed by the EDF plant is recovered at the end of the line. The cost and the additional burden of the distribution equipment needed (the network of high and low tension wires, transforming and dispatching stations, underground cables in cities, etc.) made electrical heating so expensive it could not possibly compete with other energy systems.

To get around this latter obstacle, dwellings heated by electricity were given a double advantage. They had exclusive right to an insulation that would lower their heat consumption by half, and they were given a discount for electricity which was based on typically capitalist calculations of amortization of the plants and marginal costs.

In fact, the introduction of "total electricity" can only be explained from the perspective of "total nuclearization." It was supposed to prepare the ground, the atmosphere, and the network for the atom to take over from fossil fuel. EDF saw further ahead than anyone thought.

The American Model

The nuclear changeover would nevertheless be of little interest to the French ruling class unless it was accompanied by one of the most important political and industrial turnarounds in the last 30 years. The changeover had to go along with the amalgamation of the French nuclear industry and its integration into the world strategy of the two U.S. multinationals that were manueuvering for (but didn't yet have) hegemony.

The French industrial bourgeoisie, in other words, was only interested in nuclear power insofar as the French program allowed them to be pulled along by the Americans. The latter were to take care of licensing and manufacturing specifications, guarantee the reliability of the product, and insure to their French subsidiaries subcontracted markets all over the world. In this way the French ruling class thought to protect themselves against all technological hazards and commercial risk. They were going to put French labor at the service of American brains, and then square it by giving up some of their profits to the Americans.

But for that plan to succeed, the French leaders still had to get rid of the French designed graphite-gas reactors, on the one hand, and on the other hand, to arrange the dismantlement of the CEA (Commissariat à l'Energie Atomique—Atomic Energy Commission), which had perfected this national design and which hoped to make it a winner in France and even in the rest of the world.

This was in 1968-69. It was during this time that a mission from a large South American country, which had come to purchase French reactors, was politely refused. It was during this time too that the most ordinary parts of French nuclear plants had surprising breakdowns and the semiofficial committees were formed to demonstrate the economic superiority of U.S., light-water reactors.

Then, 16 October 1969, at the time of the opening of the large (460 Mw) graphite-gas installation at Saint-Laurent-des-Eaux, Marcel Boiteux, executive director of EDF, bluntly admitted the meaning of it all. Here are the most significant quotations from his confidential speech:

> We have to acknowledge that a light-water model is not more reliable than a graphite-gas model.... But the world currently has around 80,000 Mw under construction or on order from light-water models, while there are 8000 in service or on order from graphite-gas models. You see the disproportion....
>
> For France, within our little borders, to continue pursuing a technology in which the world has no interest (sic) doesn't make sense today. The fact that the world market is now clearly oriented towards light-water models means that our industrialists will only be able to enter

the industrial world insofar as they have their own valid
experience with the models the world is interested in.[2]

The morning following this speech a technician insisted on
violating the reactor's computer program, leading to a crippling
meltdown of a fuel rod. The advocates for the U.S. design had a
free field.

The Technocratic Fait-Accompli

Up to now nuclear plants have been advertised essentially as
commodities. They are something the Americans manufacture
or will manufacture, things that can be sold, and thus things
that French capitalism has a stake in producing. How useful
these things actually are, and whether their drawbacks are
important or not, are side issues entirely. No one questioned the
external costs or the energy balance sheet of the nuclear program
any more than they questioned the usefulness of the space
program or the Concorde.

Thus the nuclear lobby had a clear field in the wake of the
"oil crisis" of October 1973, when the government was anxiously
wondering how France would pay for oil when it became four
times more expensive and how it would replace an energy
source due to run out by the beginning of the next century. In
this atmosphere the econometricians of the EDF seemed like
saviors. They had made precise calculations, to the hundredth
of a centime. These calculations showed that in the future
nuclear energy would cost half as much as oil and that if 25
years hence the French were to consume three times more
energy than in 1970, they would need a program of about two
hundred nuclear units of 1000 Mw each.

The government swam to this life-preserver held out by the
nuclear lobby. Construction of nuclear plants was to be both the
necessary condition and the engine for the industrial growth to
come. No government agency is equipped to check the EDF's
calculations, nor to submit their predictions and hypotheses to a
critical examination. And so, on 4 March 1974, the Messmer
government, without investigation or public debate, decided on
the French nuclear program under the following circumstances.

President Pompidou was already seriously ill. There was no one to oppose the Commissioner for Energy and the Minister of Industry arguing for the EDF proposal. M. Poujade, who would have been able to defend the environment, had been replaced 48 hours earlier. His successor, M. Peyrefitte, was without weapons. The Prime Minister was finally the one who decided. France would speed up construction of nuclear plants and would plunge toward the year 2000 fully electrified and fully nuclear.[3]

Nothing was left but to convince, or at least anesthetize, the public by quickly confronting it with *faits accomplis* which, despite their far reaching consequences, were always presented as technical decisions that could be competently made only by technocrats.

The Rise of Electrofascism

Throughout the past year, EDF, supported by those who are responsible for nuclear safety and protection against radio-activity, has tried to keep people from interfering in their business. The radiotoxicity of plutonium? Foolishness. Biological concentration of radioactive wastes in the food chain? Scientific impossibility. Accident risk? So far not one victim. Thermal pollution? Rubbish—people fish right below the plants at Chinon. The objections of ecologists? They are weirdos whose goal, says a confidential EDF circular, "is to impede the satisfactory functioning of today's society."

The present social order, which, as everyone knows, is satisfactory, won't be able to function without nuclear energy. And nuclear power will not be able to develop unless people have confidence in technicians and experts, who are the only true custodians of knowledge, the only trustees of the public interest, the only ones competent to make decisions. "It's pointless to waste time trying to convince professional protesters (sic)," continues the EDF circular. "We must do whatever is necessary to keep the populace from being contaminated by adverse propaganda (sic)." EDF is automatically reinventing the language and the mentality of the cops. Those who oppose nuclear energy are "internal enemies," professional subversives.

Under the wing of an exultant and reassuring propaganda campaign, whose only opposition is the ecological and/or leftist press, fundamental decisions go through like a letter in the mail. For example:

• France is to have the privilege of harboring at Tricastin on the Rhône—the biggest gaseous diffusion plant in Europe. Sixty percent of its production will be exported. The plant will consume all the power produced by four huge reactors. Tricastin I will be followed by Tricastin II and by Tricastin III: that is, at least 12,000 Mw. It is hoped that enriched uranium will bring higher and higher prices, and construction of the Tricastins means enormous amounts of capital can be invested at a profit. The climate of the Rhône valley, on the other hand, its fauna, its flora, its landscape, the health and well-being of its population, bring no profit at all. You can't sell people replacements for their pleasures and scenic spots until they have been destroyed. That's what progress is.

• Without a statement of public utility or permission to build, EDF is building at Creys-Malville the first of three breeder reactors of 1200 Mw, flanked by four refrigeration towers that in the end will plunge the whole area into a permanent fog. The "four hundred" (who are now nearly 3000) concerned scientists write:[4]

These breeder reactors are prone to accidents whose mechanisms are exactly the same as those of atomic bombs. These accidents are prudishly called "nuclear excursions," and, as far as we can calculate them in advance, they equal the explosion of a few tons of TNT. Even though the probability is small, an accident would be of unprecedented and catastrophic proportions, releasing into the atmosphere an enormous amount of radioactivity, containing in particular plutonium 239. Our question is, were the people of the Lyon area consulted or even informed about the risks to which the government or its agencies have deliberately decided to subject them, risks that the experts can neither calculate nor *a fortiori* anticipate?

The answer is no. No more than two copies of the document describing the environmental impact and the hazards of the fast breeder were printed.

Here is an example of the problem of storing radioactive waste. When no one was watching, "they" (who, exactly?) decided to make France the nuclear sewer of Europe and Japan. Starting in 1979, the factory at La Hague will be reprocessing 800 tons of irradiated fuel a year. In other words, from that date on, France will be crossed every year by hundreds and eventually by a thousand special convoys carrying lead casks of highly radioactive material. Since each convoy will take more than a day to reach its destination, there will always be several on the roads, with all the political and accident risks that this implies.

But this is only the beginning. At La Hague the irradiated fuel is dissolved. After the (always incomplete) precipitation of plutonium, uranium, and the transuranian elements, the solution contains a number of extremely radioactive wastes. To prepare it for storage, it is reconcentrated 80 times and then dumped into stainless steel casks which are coated with concrete more than a yard thick. The radioactive heating of these wastes is so intense that they must be permanently cooled and kept under continuous supervision—for seven centuries!

There is a possibility of putting these wastes into glass blocks. "But," write the "four hundred," "the effect of the temperature and the radiation on these blocks over the long term is absolutely unknown."

The American physicist Alvin Weinberg, director of the Oak Ridge National Laboratory, has talked in this relation of a "Faustian bargain." Humans must pay for access to an "inexhaustible energy source with a vow of eternal vigilance." But EDF's Boiteux sees in this analogy only a "metaphysical uneasiness." That was to be expected. Attention to the long term has always been foreign to capitalist civilization. "In the long term we will all be dead," says Keynes. Do what's immediately profitable, the rest will take care of itself.

In order to solve some short term problems we are running the risk of completely insoluble problems in the long run. One hundredth of one percent of the wastes accumulated in a century will be equivalent to the radioactive fallout of 10 thermonuclear bombs of 5 megatons each. Who can guarantee that 0.01 % of the contents will not escape from the nuclear dumps every year, or even every week?

A New Despotism

"The all-nuclear society is a society full of cops. I don't like that at all. There can't be the slightest self-management in a society based on such an energy choice," said Louis Puiseux in an interview at *La Gueule Ouverte*.[5] Bernard LaPonche, assistant secretary of the CFDT-CEA union,[6] says the same thing, which is not surprising. But now listen to Jean-Claud Leny, executive director of Framatone, the company in charge of building the pressurized water reactors (a Westinghouse licensee):

> Nuclear plants are not dangerous...if they are run by competent and strictly organized staffs with a strong sense of responsibility...If we were to install small reactors to heat individual cities, there would be this risk— their operation could be entrusted to local groups which would have them operated by more or less competent subcontractors.
>
> In my opinion it is essential that few nuclear plants be constructed, and therefore that they be large, installed on *ad hoc* sites, and controlled in a quasi-military way.[7]

There you have it. When Puiseux speaks of a "society full of cops," he is still this side of truth. Nuclear society implies the creation of a caste of militarized technicians, who obey like a medieval knighthood its own code and its own internal hierarchy, who are exempt from the common law and are invested with extensive powers of control, surveillance, and regulation.

The missions of the nuclear knighthood will include in particular: running the power plants with their two hundred reactors, training and supervising the people working in the plants, monitoring and managing the radioactive wastes stored in the plants, transporting the radioactive material and coordinating the special convoys, producing and reprocessing the fissile material, supervising the production and reprocessing plants and their personnel, monitoring and managing the final dumps that will store the wastes for centuries (hundreds of thousands of years in the case of the transuranian wastes), choosing the sites for future nuclear installations, planning the long range number of plants....

The nuclear knighthood will include tens of thousands of members and will control and supervise hundreds of thousands of civilians. It will rule as a military apparatus in the name of technical imperatives required by the nuclear megamachine.

A tendency toward despotism has always been inherent in the capitalist organization of production. The entrance of every factory could bear this inscription: "Here end democratic freedoms and human rights." The basis of this "factory despotism" (Marx's expression) is the division of labor. In order for capital to stay in charge—that is, the boss or the group of managers who represents the boss—every worker, group of workers, and shop must only produce pieces with neither use value nor market value. Only the programmed recombination of these pieces creates a usable product. And the recombination of this fragmented product of fragmented work is of course the monopoly of the managerial hierarchy.

Its power is based on this monopoly. It is the necessary intermediary between the different work skills and between the different pieces of the product. Without it, the narrow skills of the workers are worthless.

The rule of capital and the impossibility of worker power (of "self-management") are built into the basic organization of the factories. Nationalizing them changes nothing and will change nothing.

Furthermore, the functions that the managerial hierarchy assumes at the factory level, the state assumes at the societal level. Technical, economic, and territorial specialization of production means that no community, city, or region produces what it consumes or consumes what it produces. In general, an area produces things that have to be combined with or exchanged for things produced elsewhere. The government plans, coordinates, and more or less guarantees the functioning of these combinations and exchanges. As the social and territorial division of labor increases, the function of the central administration becomes more important, and its technobureaucratic power increases accordingly.

Both the industrial bourgeoisie and the public technocracy have a stake in seeing that the centralized grip of the state is as strong as possible, and that local power and autonomy are as weak as possible. Centralization of energy production and distribution, in both the technological and the geographical sense, is instrumental to an unprecedented strengthening of the central government. It makes a new despotism possible.

A Self-Devouring Machine

"Assuming all this is true," say the nuclear advocates, "what are you going to substitute for the atomic plants, without which we will suffer a reduced standard of living and more unemployment?" This question is the perfect fool's snare. In fact it is based on three implicit premises, all of which are false:

a. The first premise is that the standard of living and employment depend on increased energy consumption and on the substitution of nuclear electricity for oil. In fact it can be shown that:

• Development based on zero energy growth will lead to a greater number of jobs of all kinds than development based on increased consumption of energy. The Ford Foundation inadvertently demonstrated this in a voluminous study. These jobs, furthermore, would be more enjoyable than those that go along with big factories and complex administrative machinery;

• Contrary to the myth, nuclear energy is only cheaper than oil when that oil is used to produce electricity—*and only in that case*. As soon as it is used to replace oil in industrial furnaces or the heating systems of public buildings and private homes, an electronuclear thermal unit costs two or three times as much as one produced by oil. These calculations were done by the Institut Économique et Juridique de l'Énergie (IEJE) at Grenoble.[8] It follows from this that substituting electronuclear energy for oil would lower the standard of living.

• To raise the standard of living we must center our investment primarily on saving energy, not on producing it. Investment in energy savings requires much more labor than capital; it is a small, localized investment which creates jobs (and beyond that, it reduces environmental damage). But it is precisely because it calls for human labor rather than capital that this kind of investment doesn't interest capitalism. In short, the standard of living could be considerably raised by different and better uses of the available energy.

b. The second premise is that the only fuel that is capable of replacing oil is nuclear power. In fact, in spite of its comprehensiveness, the French nuclear program is not meant to reduce French hydrocarbon imports, but only to keep them at the present level...But the IEJE has shown that stabilization could also be obtained if we called a halt to the nuclear program and invested primarily in geothermal and solar heating.

Beyond the year 2000, harnessing solar energy not only for heating, but especially for the production of energy in small decentralized units, is not an insurmountable problem.

c. The third premise is that the nuclear program will increase the amount of available energy. In fact, a study by a group of university professors and engineers from Lyon, "Diogenes," has established that until the end of the century the French electronuclear program will consume more energy than it will produce.[9]

Is that unbelievable? No, the Diogenes group is merely taking into account the *external energy costs* of the program that the EDF economists persist in ignoring. These include the costs of the networks that distribute the electricity, the costs of the Tricastin enrichment plant, the energy costs of the nukes themselves, the costs of the new highways, the costs of the reprocessing plants, the amazingly heavy costs of teaching and research institutes....

Overall, seven plants under construction consume annually as much energy as could be produced by four plants in full operation.

"Far from resolving the energy crisis, which is the apparent justification for its adoption," writes the Diogenes group, "this program will thus continue and even exaggerate it. A self-devouring monster, growing for its own sake, and artificially inflating the Gross Energy Product, electronuclear energy is the crowning achievement of a society that has become more and more complex, and more and more frenetic, but which offers less and less to the individual."

The Alternative

But you may be objecting to all this; for if it were true, wouldn't the business economists have been aware of it long ago? But maybe many of them are aware? Why should they protest as long as the nuclear program, however uneconomic, is profitable to big business? The economic calculations of the Diogenes group merely established that the nuclear program would not increase the net amount of energy available for anything besides the production of power plants.

But this is not an unusual paradox. All the richest opportunities of advanced capitalism consist in consuming and destroying free resources in order to reproduce them by complicated means and resell them to people in the guise of goods and services. And when, for one reason or another, the expansion of the market is blocked at the consumer level, capitalism arranges to have the government consume a special kind of merchandise, whose only purpose is its own self-destruction. This special merchandise, which is very profitable for industry, is armaments.

In many respects the nuclear program serves the same purpose as an arms program. It keeps capital circulating and enables it to make profits to the detriment of everyone.

The development of light technologies relying on geothermal and solar energy would have an entirely different economic nature, and are thus of no interest to capital. For investment would be decentralized, and the technology could be learned and used by even small communities or individuals. There would be no need to transport energy (especially solar energy), and large units would have no advantage at all over small ones. Thus no firm, no bank, no government body would be able to monopolize these technologies. They would give local groups and not-yet-industrialized nations a high degree of independence, and they would make a completely different kind of development possible.

This is the "alternative" that capitalism fights with all its might—at the level of multinational firms and national governments. To refuse the nuclear program is to refuse the logic of capitalism and the power of its state.

Le Sauvage, April 1975

3. Boundless Imperialism: The Multinationals[1]

The development of the multinationals puts just about everything into question: our ideas of government, power, currency, planning, nationalization, the workers' struggle, foreign trade. Everything that politicians continue to put behind these words has been overtaken. Since the 18th century the world has scarcely known a series of upheavals of such scope.

But if you consult the hundred or so books that in the course of a year have been devoted to the multinationals, you will scarcely find these fundamental aspects discussed at all. Neither will you find them in the "exhaustive" study of several thousands of pages that the Ford Foundation and IBM ordered from Harvard University. The board rooms, you see, are in the process of putting together an operation that has already succeeded with the urban crisis, growth, pollution, and "the quality of life"; this manoeuver is called "distract the opposition." From its foundations, business schools, and elite universities the multinationals order rafts of knowledgeable and jargon-ridden literature to drown out the voices of us unionists and workers, who were the first to flush this hare and who call a spade a spade. That's what they call "raising the level of debate."

They put the concept "pollution" in place of the tangible reality, which is the fact that *their* chemical, metallurgical, and oil industries are poisoning first of all those who work there, then the inhabitants of the cities, and finally the fish in the sea. In the same way they are trying to put the concept *"the multinational firms"* in place of this other reality: Michelin or General Motors, IBM or Saint-Gobain, are organizing operations on a global scale, placing and moving their factories like pieces on a chessboard, and emptying all substance from the independent policies of nation-states and governments on industrial, commercial, fiscal, and monetary matters!

The man who says this is Charles Levinson. A Canadian, secretary general of the International Chemical Federation (ICF), and author of two well-known books,[2] he first exposed the multinationals when Americans—mostly the workers—reacted to a flood of Japanese transistor radios in a spasm of nationalism and protectionism. Levinson, then assistant secretary general of the International Federation of Metalworkers, revealed to them a shocking fact. Products *Made in Japan* were in fact made by Japanese subsidiaries of U.S. companies. The invasion of the American market was being directed from New York *via* Tokyo. The enemy of American workers was neither the Japanese worker nor even the Japanese government; it was U.S. capital itself, which was American only in origin and name.

Twenty years have passed since then. And what was then sensational is now quite ordinary. Rolleiflex cameras are put together in Singapore, ditto Siemens microcircuits. Agfa-Gevaert manufactures its cameras in Japan. "Swedish" furniture comes from Polish factories. Some Renault autoparts are made in Yugoslavia and Rumania, etc. As for the U.S. companies, many of them do all their manufacturing abroad. All cameras sold in the United States are made abroad. The same goes for 96% of all tape-recorders, 95% of all two-wheelers, 90% of all radios, 70% of all portable typewriters, 67% of all shoes, 50% of all black and white TV sets, etc.

At the present time there is no large firm that does not own subsidiary factories in several countries and that doesn't entrust them with manufacturing various parts or components of its products. Presently a firm's ability to survive as an independent entity depends—and will depend more and more—on the number and stability of the subsidiaries it has established around the world. Later on we will see better the advantage this "multinationality" offers. For the moment, listen to Levinson again as he describes the scope of the "multinationalization" process:

Multinationals' production is now growing twice as fast as the whole of world economic activity. Certain writers are convinced that by 1985 two hundred to three hundred corporations will control 80% of the West's productive assets. If you look at "advanced" industry—which is called "scientific" because it uses a little labor and a lot of grey matter—you will note that a handful of firms, often associated in consortiums or joint ventures, already dominate the world. Seven giant firms control the whole oil industry. Fifteen control petrochemicals. The electronics industry is controlled by ten firms. The tire industry is monopolized by eight companies, the manufacturing of plate glass by five, paper production by nine, etc.

And if you think these giants are fighting each other tooth and nail to enlarge their respective shares of the market, disabuse yourself right away. There are certainly cases and places where competition is still alive, but among well-established firms the tendency is no longer to battle with each other, but to have cartel agreements—gentlemen's agreements—to help each other consolidate control and bar the way to newcomers.

Take the tire case. You've been told that Michelin has had problems in North America, where it is trying to establish large factories. You thus conclude that there must be a lively battle among the tire giants: Dunlop-Pirelli, Michelin, Goodrich, Firestone, and Goodyear. But then you discover that in several countries Dunlop makes tires for Goodyear, that Michelin and Dunlop are partners in joint ventures, and, to top it off, that an *Irish* manufacturer who produces radial tires for an *American* firm belongs to the *Austrian* firm Semperit, which is controlled by the *Franco-Belgian* firm Kléber-Colombes, which itself is controlled by the French firm Michelin, whose head office is in Basel (Switzerland).

So when people talk about the *battle of the giants*, take it with a grain of salt. The *real* giants don't fight among themselves; there are too many risks. Their differences are worked out around the conference table. This is how Shell comes to be involved in 25 joint ventures and Standard Oil of New Jersey in 35 joint ventures with other oil companies. And how it is that there are more than 4000 joint ventures among American multinationals and their supposed European competitors—three times more joint ventures than there are among European firms themselves.

The picture that emerges from these facts is one of a world oligarchy, made up of a few hundred large firms whose managers (including, as we shall see, the new Soviet managers) come from the same schools and the same social class, have the same ideas, and pursue the same ends by the same means. These firms certainly compete for the markets, but they never resort to commercial warfare or cutthroat competition. The weapons they use against one another are rather the novelty and the "image" of their products—created and maintained by expensive advertising campaigns—or the expansion of their commercial networks, in particular by constant efforts to seduce the retailers and salesforce of other brands. In addition, they use bribery, which is effective, lucrative, and in fact indispensible wherever the state and its agencies are corruptible...which is to say, everywhere. In Europe, while you cannot as a general rule buy the good will of the heads of governments, you can always buy or rent the services of ministers, former ministers, or political figures who have one claim or another on the ministers.

"Don't forget," says Levinson, "that the eminent 'European' Paul-Henri Spaak, a former socialist premier of Belgium, was at the time of his death director of ITT-Europe, and that another great 'European,' Louis Armand, former president of the state owned French railways, was director general of Westinghouse-Europe."

But all this is really only a small part of the picture. One begins to perceive the root of the problem when one looks at the weight of the multinationals on the foreign trade balance, the balance of payments, and the monetary policies of governments. Most people, including politicians, still think that the foreign accounts of a nation resemble those of a grocery store—exports on one side, imports on the other. If the country is importing more than it is exporting, it should devalue—that is, lower its export prices so that it can sell more abroad. Well, no. Things don't happen like that at all anymore. A modern capitalist economy has by definition a commercial deficit with foreign countries. If, like Germany, it has a chronic export surplus, that's because it is not yet quite modern. Levinson has shed light on this in a provocative way.

The production of the foreign subsidiaries of the multi-nationals already exceeds by more than 20 billion dollars the total amount of world exports. The exact figures for 1971 are: total world exports, 310 billion dollars; production of foreign subsidiaries, 330 billion dollars. And of these 330 billion, 275 were the exclusive achievement of American subsidiaries. So you can say that these latter account for a production equal to 90% of the total amount of exports of all countries taken together.

Thus, to worry about the American trade balance, to think that the situation can be corrected by monetary manipulations, is pure humbug! There is practically no large American firm that cares about exporting its product. The American firm exports its capital, its factories, its know-how, and its trade networks by directly establishing its subsidiaries in the country whose market it wants to take over. Currently U.S. capitalism produces abroad six times as much as it exports. And this ratio will rise to eight in 1975. Two-thirds of U.S. industrial exports are made up of products and services that the domestic firms *sell to their own foreign subsidiaries.* To all intents and purposes the bulk of U.S. exports is the result of capital export. And if you add to this fact that the subsidiaries are captive clients, to whom the head office can sell at absolutely fantastic prices, you are obliged to conclude that monetary manipulations would have only a very limited effect on U.S. exports.

Here Levinson gets to one of his favorite themes: the external deficit of advanced capitalist economies is a strategic necessity. In his last book he wrote, "the export of material goods is already outmoded; today it is management and capital which cross frontiers without concern for customs or other barriers, which do not affect them anyway." As early as 1916 Lenin observed: "The export of goods that characterized old style capitalism is giving way under modern capitalism to the export of capital." The examples of Germany and Japan make it easy to understand why.

Ravaged by the war and deprived of colonies, Germany and Japan have had since 1948 rates of investment unequalled anywhere. More than a third of their national product was invested. Wages were much lower than in other industrialized countries so as to make possible this record accumulation and growth. Gigantic industrial complexes arose with a capacity that quickly exceeded the needs of the country.

Industrialists had no choice. Too weak politically to carve out new spheres of influence in a world dominated by their "conquerors," they could only invest in their own countries. And that is what they did. The major firms thus grew enormous. To make their establishments profitable, they set themselves to export 50 to 60% (and often more) of their production. And since, because of their low wages (low in relation to productivity), the Germans and Japanese could not consume internally imports equivalent to the exports, their countries found themselves earning year after year enormous amounts of foreign currency for which they had no use. Most other countries found themselves in debt to them.

In the opinion of the German and Japanese governments, things couldn't go on this way. Indeed what could they do with these mountains of hard-to-convert currency? And how can you continue to sell to other countries if they owe you a lot of money already and have nothing to sell you that you need to buy? Wouldn't it be better for Japan and Germany to build profitable factories in the customer countries rather than to sell them German and Japanese merchandise on credit?

This question, first raised by the most active companies, finally became an extremely important one for three reasons:

1. As a result of overindustrialization and the labor shortages it leads to, wages "exploded." Immigrant labor was called in. But in its turn this labor becomes less and less tractable. Furthermore, imported labor can't be indefinitely penned up in the shanty-towns that might satisfy them if they stayed in Africa or Turkey or Korea or the Philippines. So you might as well build the factories right in those countries and benefit from the low wages. This reasoning applies mostly to manufacturing industries (automobiles, precision tools, optics, television sets) as this statement of Ernst von Siemens, head of Siemens, shows. It is reported by

Levinson.

> Half of Siemens' sales were made in foreign countries, even though only 20 percent of its production occured outside Germany. In the future we are going to export more capital and know-how instead of continuing to import foreign workers—who are already 20 percent of the workforce in Germany. In the course of the next decade, personnel hired outside Germany will increase by 50 percent, and only by 10 percent within Germany.

2. Overindustrialization, notably in Germany, Japan, and Holland, is running into physical limitations. Space, air, and water are beginning to be in short supply; overpopulation and air pollution in industrial centers have passed the critical point. This is the main reason why the German chemical industry cannot build new factories along the Rhine. They'd have to recycle the water and air, and build new cities. It's cheaper to build subsidiaries in Brazil and in the southern United States— along with capital, industry exports pollution.

3. These two reasons aside, foreign subsidiaries offer such great financial and political advantages that all large firms owe it to themselves to consider them. "We can estimate," says Levinson, "that a multinational company makes a 30 to 40% higher profit than a traditional export company." This is particularly true thanks to the transfer price mechanism.

Explanation: a multinational company, Michelin for example, has 17 factories in 13 countries. Atop these factories there is a holding company which is in charge of financial management, and a commercial company which regulates imports and exports. These companies are domiciled in Switzerland (in Basel, in Michelin's case) or in one of the "fiscal paradises" where there are no taxes on profits—Liechtenstein, Luxembourg, Cayman, the Bahamas, etc. International managers will then see to it that the subsidiaries make no profits, for example, in France, in Holland, or in Germany where taxes are high, but that profits will be all the more substantial in countries where taxes are low or nonexistent.

To achieve this result, the international management of the firm charges its subsidiaries disproportionate prices for "services rendered," manufacturing licenses, spare parts, semi-finished products. For example, there is nothing to keep the "Swiss" management of an automobile company from having its parts made in Poland, and then selling these parts to its French or German factories at a profit of 200 to 300%. By such varied means the Swiss-based international management realizes gigantic tax-exempt profits at the expense of the French or Belgian factories, which, since they are artificially burdened by staggering costs, will always appear to be on the edge of deficit.

By choosing residence in a fiscal paradise, the multinational firm enjoys the following advantages: it pays no taxes on profits; it needn't worry about credit restrictions or the control of foreign exchange that might be established by one or another of the countries where it has factories; it reduces the risk of nationalization or expropriation. Indeed, if a French government nationalizes Michelin, Pechiney, or Saint-Gobain, for example, it will only take over the French factories. The foreign subsidiaries, property of "Swiss" holding companies, will remain out of reach and in retaliation may conspire to cut off the French plants from their sources of supply and their foreign outlets.

But all these practices are still relatively mild compared to the systematic exploitation of the Third World countries. There, neither competition nor political power curbs the large firm's thirst for profit. It descends on a practically untouched market, starts out by buying the goodwill of the Interior Minister, the chief of police, highranking officers, and local dignitaries, and then it sells the goods and services of its subsidiaries at exorbitant prices. For example, in Brazil, the British or Swiss pharmaceutical industry sets prices that are sheer robbery. Railroad and telecommunication prices in Central America, which are under North American control, are the highest in the world.

However, for the past twenty years the profits realized in this way have become more difficult to take out—the indebtedness of the pillaged countries has reached a ceiling and their currency is difficult to convert. So the American and European companies have changed their methods. Now they take their profits not only from the sales made by their subsidiaries, but also and especially from the supplies that the firm's headquarters or home office sells to them.

For example, the head office sends the subsidiaries in Argentina or India already paid-off machinery and makes them pay up to four times the usual price. They licence their patents to them and sell them "managerial" services at robber prices. They force them to use raw materials and components supplied exclusively by the home office at monopoly prices. In sum, the subsidiary becomes a captive client of the firm's head office, which often realizes the bulk of its profits not from the merchandise its subsidiaries sell, but from the goods which it forces its subsidiaries to buy.

According to a recent UN report, the Third World subsidiaries pay a sum for patents, licences, and "services rendered" that is equal to half the new investments going into "underdeveloped" countries. The superprofits that the multinationals realize on supplies they ship to their subsidiaries is undoubtedly equal to at least the above mentioned sum. These concealed superprofits are of course not included in the officially stated profit rates of the multinationals in the Third World. These profit rates are nevertheless impressive: non-oil investments officially bring in 12% of the capital invested in the southern hemisphere, as against 10% in Europe and 8.6% in Canada.

Such are the principal advantages a firm gets from the multiplication of its subsidiaries. In the final analysis, a firm's "multinational" operations are nothing but what five years ago was called "economic imperialism," or simply "neo-imperialism." There was a lot of noise about this in France during the 1960s. At that time North American firms, still strong in their technological supremacy and universally coveted dollars, invested so heavily abroad that in 12 years (1960-1971) their international wealth went from 32 to 86 billion dollars. This means that in a 12 year period the foreign investments of U.S. firms were *one and a half times more than during all previous history.*

A strongly rooted myth suggests that these direct investments are the major cause of the U.S. balance of payments deficit. The reality is much less simple. The most important of the new U.S. investments in Europe were financed by European banks and governments themselves, which were delighted to lend money to a transatlantic firm or to give it public subsidies in order to attract it to Bordeaux, Dunkirk, Rotterdam, or Bavaria. And what's more, the U.S. firms had no need to go into debt to finance their foreign investments. During the three years 1968-1970, for example, they repatriated (mostly from the Third World) a total of 24.3 billion dollars in dividends, interest, royalties, and various repayments. During the same period they invested only 10.9 billion dollars abroad.

In 1971, the last year for which complete figures are available, U.S. firms officially repatriated 9 billion dollars in dividends, interest, royalties, and repayments (to be precise, 6.67 billion in dividends and interest, the rest by way of royalties and repayments). In this same year they only invested 4.8 billion dollars of new capital in their subsidiaries.

Again, these figures don't tell the whole story. They are only the tip of the iceberg. In fact, a multinational only repatriates those profits to the degree that it cannot profitably reinvest them abroad, either because the market is too small to absorb further production in a particular country, or because the political risks of new investment are too great. And that's generally the case in most of the Third World countries.

In addition, the multinational firms take the bulk of their profits from the underdeveloped world—this was flagrant in the case of Chile in the 1960s—and place them in prosperous and politically stable countries and regions like Canada, Western Europe, and Australia. The UN figures speak eloquently about this. During 1970 the multinationals took 996 million dollars out of Africa, but only invested 270 million there; they took 2400 million dollars out of Asia, but only invested 200 million there. Repatriated money from Latin America rose (from 1968) to 2900 million as against 900 million in investments. Again, these figures don't take into account the concealed repatriation, which for Latin America is equal in size to the open repatriation.

In the light of these facts, imperialism, the "pillage of the Third World," becomes a tangible and statistical reality. U.S. capital is effectively exploiting the rest of the world. And according to an expert's report done for the Rand Corporation, that's only the beginning. Before the end of the century, this report predicts, the United States will have all their manufacturing done abroad and will have on their own soil scientific and service industries. If perhaps you are wondering how they will pay for the manufactured goods they import, since they will no longer export any commodities, the answer is: they will pay for them with the profits brought in from U.S. factories all over the world.

Americans, according to this forecast, will become a people of bank employees, of technologists, and of soldiers, principally concerned to protect and multiply the billions raised through the toil of other peoples. Thus in the world of the 21st century they will be a superpower comparable to Great Britain in the 19th century. On the condition, of course, that they don't stumble over either "more Vietnams" or a large federation of countries that rebels against U.S. control, in the way that the United States once freed itself from British control.

Is is conceivable that "Europe" might one day be this federation of countries? Doesn't it compete with the United States in different parts of the world? Doesn't it, like the United States, hope to live on the revenue of its foreign investments? Doesn't it already control 41.7% of the total capital invested abroad, as against 52% for the United States (1971 figure)? And isn't the U.S. losing ground, since in 1967 it controlled 55% of the total foreign investments while Europe's share was only 40.3%? Doesn't Europe include old imperial countries like Great Britain, with 24 billion dollars in foreign investment (14.5% of the total), France, with 9.5 billion (5.8% of the total), Germany, 7.3 billion, and Holland, 3.6 billion?

Slow down. Reality is less propitious for "Europe" than these perceptions and recollections make it appear. For "Europe" hasn't the military and political means for a global strategy; and the quality of its investments is not up to those of the United States. One fact will make this clear. The amount of European capital invested in the United States is about equal to the amount of American capital invested in Europe. But American capital controls entire branches of industries that have strategic importance. European capital in the U.S. controls nothing. It is invested in U.S. stocks and bonds, and it stays in a junior position.

For 50 years the two main imperial countries in Europe, Great Britain and France, have not had an industrial base that could compete with U.S. expansionism and impose their own world policy. French foreign investments are essentially in mines and concentrated in Africa. Great Britain's are essentially financial and commercial. Holland's (thanks to Philips) and Switzerland's (Nestle, Brown-Boveri, Hoffman-LaRoche, Sandoz, etc.) are very modern in structure and rely on advanced technology. But Holland and Switzerland have no political or military weight and therefore cannot protect the interests of their multinational firms or intervene in the politics of the countries where those firms have investments. Thus Holland and Switzerland are aware of a permanent need for the military and political protection of their worldwide interests, and they willingly follow the imperialist policies of the United States—which alone can keep "law and order" in the rest of the world. This goes a long way toward explaining Dutch "Atlantism" and the ultra-conservatism of Swiss foreign policy, as well as the pro-American nature of British policy.

So is the European-American rivalry just a dream? Yes and no. It's a reality insofar as European capitalism—mostly French and German—is redoubling its efforts to cut out for itself a multinational industrial empire that resembles the U.S. empire. Germany's foreign investments have been increasing at an annual rate of 23% since 1960; France's doubled in 1971 and further increased by 35% in 1972. A considerable part of this investment is carried out by young firms. These firms don't yet have the financial size and power that would make them comparable to the real giants, American or not. They have to strengthen their hold, first over the national market, then over the European market. They ask to be protected against invasion by U.S. subsidiaries not only on their own soil but in the whole of the Common Market. They are asking for "European" customs barriers and a "European" policy to regulate U.S. investments.

On the whole they want Europe to remain their preserve until the day when they are strong enough to launch their own conquest of the overseas markets. And when that day comes, Europe will not be large enough for their ambition. Their European "nationalism" will have lost its raison d'être, just as it has for such "European" giants as Philips, Fiat, Saint-Gobain, Péchiney, l'Oréal, Michelin, BASF, ICI, Volkswagen, etc.

This analysis is set forth by three economists—Bernard

Jaumont, Daniel Lenegre, and Michel Rocard—in their book *Le Marché Commun Contre l'Europe*.[3] Are we to infer that European corporate capital has no interest in the creation of a supranational state and institutions? The three authors think this is the case. But the question is a controversial one, and other left economists (Ernest Mandel and Robert Rowthorne in particular) present more shaded analyses.[4]

For, aren't "European" multinational firms more "Atlantist" than "Europeanist" for the sole reason that only the Pentagon and the CIA are able to defend the capitalist order from Cape Verde to Mozambique, from the Philippines to Suez, from Alaska to Tierra del Fuego? Won't these big European firms feel the need for their own supranational state and politico-military instrument in order to contend with the United States for Arab oil, Siberian gas, and the Argentinian and South Asian markets? And, at last, when there are only two or three giant "European" corporations left in electronuclear energy, aeronautics, and electronics, won't they—in order to assert themselves in the rest of the world, and even to safeguard their independence—ask that a European state authority establish and finance "European" investment programs, "European" export policies, and public subsidies without which European companies would be beaten before they begin by American firms—which are not short of public subsidies (in the form of orders and contracts for military research)?

All these questions lead to a new interrogation. In the years to come will there be a world crisis (that is, a generalized recession)? If so, will it be the occasion for a confrontation between Europe and the United States and for the creation of a European state? Or will European institutions be merely the instruments that allow the big firms of Europe to pursue a global strategy in association with the U.S. firms, ending up with the unification of world capitalism under U.S. hegemony?

The bets are not all in. But right now time is not on Europe's side. As Levinson showed in the case of joint ventures, as far as the big firms are concerned, the integration of Europe and the United States is advancing more rapidly than intra-European integration. The European governments that are encouraging this "transatlantic" integration of businesses and banks persuade themselves for better or worse that the issue of national sovereignty, or even European sovereignty, is obsolete. When the government is defending the interests of "our" big firms, it is no longer defending the national interest, but rather a capitalism that has neither nationality nor country. The three authors of *Le Marché Commun Contre L'Europe* put it very well:

> Imperialism in the classical sense will be less and less embodied in the governments of the capitalist world. Their only function will be to insure that conditions on their national soil are satisfactory to all business enterprises, regardless of their nationality. Every government will become a spokesperson for the big firms, no matter where they come from. These firms do not need a world government for their interests to prevail. All that is needed is for the secular power of the capitalist community in every country to intervene against those who threaten the interests of capitalism in general... The French government in New Caledonia or the Portuguese government in Angola does not defend the interests of French or Portuguese firms alone. They are both acting as watchdogs for capitalism in general and simply keeping anyone from harming the profit economy.

Levinson pushes the analysis even further. In the East, as in the West, he thinks governments have become autocratic machines controlled by "elites" who are indistinguishable from each other and who help each other perpetuate their own power. To show how this works Levinson first of all tells this true anecdote. An Austrian glassworks factory was having problems with its joint production committee, most of whose members belonged to the Austrian Communist Party. What did the boss do? He decided to dismiss his 200 workers and move his factory to Hungary, where wages are lower and where he wouldn't have to be afraid of strikes. The president of the Austrian chemical workers union went to Hungary especially to ask for brotherly assistance from his Hungarian counterpart. They answered: "*This is an economic problem which is none of the union's business.*" Levinson continues:

I could tell you many other stories of cooperation between our capitalists on one side and the Eastern managers and bureaucrats on the other. More than 900 firms now have investments in the East. The Italian firm Montedison is announcing that it will invest 500 million dollars over a 20 year period. Armand Hammer, president of Occidental Petroleum, has signed an eight billion dollar contract with the Soviets for the creation of fertilizer factories. To do this, Mr. Hammer obtained a wholesale loan from the same Export-Import Bank that refused to loan anything to Allende's government.

And, to crown the irony, when Brezhnev and Chelepine, during Allende's visit to Moscow, promised him their brotherly support against the ITT conspirators, they had already signed a contract with ITT for several million dollars for (among other things) equipment for Soviet airports.

Mr. Rockefeller, of Standard Oil and Chase Manhattan Bank, has already made 700 million dollars available to the Soviets. His cousin, of the First National City Bank, has agreed to a similar amount. I call this political pederasty. You attack capitalism but embrace the capitalists who themselves are doing the same thing with socialism and socialists. The elite of both camps now have the same ideology, which is roughly that of the Harvard Business School—whose disciples are currently organizing courses in Moscow. Thus we shall have the same methods of management, the same hierarchical division of labor, and the same military discipline in the factories of Detroit and Togliattigrad, of Chicago and Minsk. American managers are teaching the Soviet bureaucracy how to introduce the profit system without weakening its power. Conversely, on the backs of the Western working class, the Soviet bureaucracy is helping the capitalists overcome their difficulties.

Let's use our heads; where do these billions that are going to be invested in the USSR come from? Marxism has told us well enough: they are the surplus value accumulated from the exploitation of American and European workers. In short, socialism now wants to build on the exploitation of *our* workers by making deals with *our* capitalists.

Levinson continues:

But here is something even more serious. These giant enterprises that our capitalists are establishing in the East will operate on the basis of co-production. This means that the American or European firms will obtain at cost a fixed share of the fertilizer, the plastics, the tires, and the cars produced in the USSR. They will sell this share in the West at market prices—that is, at a substantial profit. In addition to goods from the subsidiaries in the Third World, we will be getting those from American-Soviet, Italo-Polish, and Franco-Rumanian companies.

And what if our workers go on strike? You can see the problem: After 20 years of work, our international unions are in a position to mobilize in Ohio and the Philippines, in Germany, in Venezuela, and in Switzerland, the workers who work for the same company. For example, we have just recently succeeded in keeping some Spanish workers who occupied their factory from being dismissed and indicted. Such united action could keep a subsidiary in Holland from being closed down, or a French workers' strike from being balanced out by intensified work in Belgian and German subsidiaries. But the hybrid enterprises of the Eastern countries threaten to sabotage this work. There are no independent unions in the East; the right to strike doesn't exist. The company manager belongs to the same party as the union head and makes sure that the union drives production forward. Thus, whenever we want to enforce our demands by a multinational action, our unions will have against them, in addition to our own managers and our own governments, the managers, the governments, and the unions of the Eastern countries.

What I'm explaining here are not possibilities: they are certainties. During a recent visit to the USSR a delegation from the Italian CGIL asked representatives of the Russian workers' union: "If Fiat struck in Italy, would we be able to count on the solidarity of comrades in the Soviet Fiat plant?" Answer: "These are political questions. Here the union doesn't meddle with politics."

Levinson continues:

So from now on proletarian internationalism is limited to the capitalist world. The FSM has lost any possibility of putting into practice an international strategy for the workers' movement. Here is the new situation. What can we do with it? I will tell you. The workers' fight against hierarchy and for workers' power must be waged in the East as well as in the West. In order to accomplish this we must be able to make contact with those who truly represent the workers of the East. The spirit and the methods of the workers' struggle must now be spread from West to East and no longer the other way.

We couldn't do a bigger favor for the Soviet economy and people. For one thing is certain: American management methods are not efficient in themselves. What makes them efficient is constant pressure from the unions and the refusal of the workers to accept just any old thing. This compels the managers to get smart and to keep inventing new machines. If the Soviets take our managers, they must also take our unions. Otherwise, they will experience the same kind of disaster as countries that combine American style management with fascist politics.

15 October 1973

4. Labor and the "Quality of Life"[1]

The General Context

In its heartland, as well as on its periphery, the capitalist world is entering a new period of upheavals and crises which will probably spread to most aspects of our way of life in the next decades. Continuation of the kind of development we are familiar with is going to run into (and has already run into) limits that are as much external as internal. The higher cost of the principal factors of production, the slower rate of technological innovations, the emergence of physical bottlenecks, and the increasing importance of transnational trusts will make the system more and more inflexible. In such a state of affairs the traditional bargaining methods and objectives of the labor movement will come up against structural and political opposition within the system and will be much less likely than in the past to wrest from it any improvements in working conditions.

Before looking at the new tasks and problems this turn of events will create for the labor movement, let us try to describe the general context in which the coming struggles will unfold:

1. The causes of the predictable rise in cost of the factors of production will be as much political as physical.

They will be political insofar as the price of raw materials has so far been tied to the imperialist domination and pillage of Africa, Asia, and Latin America. This domination has little chance of continuing to the end of the century, and its defeat will inevitably be accompanied by a sharp rise in the cost of raw materials.

They will be physical insofar as depletion of abundant and easily accessible deposits of a whole group of mineral resources, and the need to work deposits that are poorer and harder to get at, will, according to experts' predictions, drive up the cost of nine indispensible metals by a factor of ten.

2. The rise in the costs of reproduction will be due mainly to the need to maintain and to *reproduce* the natural environment. In other words, diseconomies and external damage—which capitalist growth has so far disregarded—will from now on have to be taken into account in any calculation of production costs.

From this point of view it is a mistake to think that repair or reproduction of the natural environment, in particular fighting pollution, could propel or maintain the growth of the capitalist economy as a whole. Even though the necessary investments could be sources of profit for some particular capital (Teil-kapitale), they add to the reproduction costs of aggregate capital (Gesamtkapital); and they will weigh down the profit rates and/or will raise the prices of consumer goods.

From another angle, the need to preserve the natural environment and the increasingly scarce mineral resources will force the developed capitalist world to revise or abandon a *consumption model* that is based on an artificial stimulation of needs, obsolescence, and an accelerated turnover of goods—a policy which, as Barry Commoner has shown, is a major cause of the destruction of nature.[2] However, a slowdown in the rate of obsolescence and greater product durability will correspondingly slow down the turnover of capital and will be an additional cause of falling profit rates.

3. The exhaustion of growth factors that prevailed during the past three decades, predicted since the early 1960s by economists like Ernest Mandel,[3] is beginning to be evident. This refers to both the saturation of the "durable goods" market in the major capitalist countries and the levelling off of the effect that the technoscientific breakthroughs of the 1940s and 1950s had on productivity and on product innovation.

New Fields of Action

The period now beginning will inevitably be marked by stagnation or the slowing down of growth, the contraction of employment, and the shrinking of the economic surplus that enables the capitalist system to finance large-scale reforms and large social programs. When the system becomes this inflexible and vulnerable, the traditional division between political and economic (labor) struggles tends to lapse. There are three main reasons for this.

1. The usual immediate demands—wages, hours, working conditions, job security—come up against increased resistance and are considered by capital as attacks on the basic equilibrium, the stability, and even the viability of the system. This *objective* politicization of the labor struggle has been seen recently in the United States and in West Germany, for example, while in Italy and France the labor movement is openly beginning to draw the logical conclusions from this situation. Since all union action in fact assumes a political character and comes up against retaliations and counterattacks that are political in character, economic struggle needs to be linked to the *conscious* rejection of capitalist logic and the *conscious* attempt to change society. This develop-

ment is particularly apparent in the case of the French CFDT (Confederation Française Democratique du Travail) and the Italian FIM-CISL (the non-Communist metalworkers' union), whose tendency is to bypass unionism in favor of building a mass political movement. I shall return to this.

2. While they are still of fundamental importance, traditional economic demands no longer fully account for workers' demands. These seem more and more often to be tied to extra-economic "qualitative" goals, which call into question government policies, the employer's power and prerogatives, the organization of work, hierarchy, lifestyle, etc. A poll taken at two big Régie Renault plants (at Billancourt and Le Mans), for example, showed that 56 and 85% of the workers, respectively, thought reduction of working hours was more important than a wage increase. In 1967 only 31% of the workers indicated that preference. In four years these workers had discovered the basic difference between earning more and living better. Independently of English-speaking economists like Galbraith[4] and Mishan[5] they came to the conclusion that expanding production (Gross National Product) had stopped improving their living conditions, and that higher wages alone would not insure them a better life. Living better depends less and less on individual consumer goods the worker can buy on the market, and more and more on social investments to fight dirt, noise, inadequate housing, crowding on public transportation, and the oppressive and repressive nature of working life. A text from the CFDT bears this out.

> The CFDT considers that in the current stage of capitalist development the condition of the workers is more and more shaped by their existence outside the workplace, by the framework of their lives (transportation, housing, environment, etc.), and by news, culture, teaching, health, consumption, leisure activities, etc. Industrial capitalism shapes these various areas in such a way as to make sure that the people's tastes, behavior, culture and dreams will be instrumental to the smooth functioning of the system, and to the growth of new markets.

> Whenever capital has to concede higher wages, it tends to make up for them by slashing expenses for public

services and collective facilities—except obviously those that are necessary as infrastructure or for economic development from the capitalist point of view. The recent struggles over transportation, living space, pollution, health, free time, education thus become more and more decisive insofar as a large part of the living standard and lifestyle are at stake in these areas.[6]

In characteristic fashion, this document insists on the need to go beyond wage demands and the business sphere in order to put the workers' needs into the large perspective of a plan for civilization—that is, into the perspective of the autonomous definition and satisfaction of needs and aspirations independent of the capitalist market. Having abandoned any illusions as to the possibility of getting this satisfaction through reforms or reoganization within the capitalist system, the authors take a revolutionary and socialist line that deliberately erases the traditional division between the work of the union and that of political parties.[7] "Always building on the actual experience of the workers, the CFDT's strategy is characterized by a constant readiness to use any situation to launch a massive onslaught against one or another characteristic of capitalism—whenever the people seem ready for it. The CFDT endeavors to develop socialist consciousness and autonomy of thought and action in the workers, so as to enable them to be the main agents of social change and the source of all economic and political power."

3. More fundamentally, the commodification of all spheres of activity, the mercantilization of all rights and resources, and the concentration of power in oligopolies that overlap with governmental power have led to the decay of civil society, to the disintegration of the social fabric, and to an irreversible crisis of bourgeois ideology. The self-perpetuation of the state is no longer based on its capacity to win a consensus, nor on the popular appeal of its ideals and goals, but on the tricks and corrupt dealings of unseen forces, on the bureaucratic power of centralized systems such as the administrative agencies, the police, the army, and, often, the unions.

This disintegration of society—of which the United States and Japan offer the most striking examples—is characteristic of a pre-revolutionary period. The state, progressively shrinking into an administrative machine, can no longer govern and limits itself to crisis management, thereby risking a relapse into authoritarianism and barbarism. The discrediting of traditional political parties, of party politics, and of electoral and parliamentary intrigue implies a new challenge for the labor movement: that of building a novel type of political force whose mass appeal and democratic functioning would foreshadow the downfall of the old order and the advent of self-government by the people.

What To Fight For

I shall discuss further on the possibility and the limits of a similar transformation of the union. What is to be emphasized for the moment is that the widening of the union's sphere of activity and the working out of an overall political-ideological concept can not be simply a reaction to the increased rigidity of the capitalist system, but must offer a common ground for action to a highly differentiated class of manual, technical and intellectual workers. For their unity in action can never be obtained by adding up the immediate interests of their respective trades, but only through an overall vision transcending these interests. Today's working class is too highly differentiated for its unity to have an immediate material basis. Its unification will have to be *constructed* by systematically attacking the roots of division from a class perspective.

When it doesn't make this kind of effort, the union tends to lose the ability to lead, channel, and coordinate the struggles. It becomes a reflection of the disintegration of the working class, and is itself threatened with disintegration, as in the case of some of the British and U.S. unions. When a union is constantly surprised by wildcat strikes and by local and job-specific movements whose motives it was unable to sense in advance and interpret so as to mobilize the workers around their own issues, it loses control and ends up being afraid that these "uncontrolled" movements will jeopardize its bargaining power. It ends up in the role of firemen—rushing in after the fire has already broken out to try to put it out. This has been the general situation in the developed capitalist countries for several years. In France the recent development of long, hard, plant-specific struggles shows primarily that worker groups are rebelling against the methods and objectives of the national unions. These "wildcat strikes" are specific not because of their narrow-mindedness, but because the rank and file cannot successfully go beyond the local and plant-specific level as long as it has to organize its battles *without* and often *against* the union.

A few Italian unions have made the most significant attempt to win back the initiative (and also control) of the struggles by expressing issues that anticipate and deepen workers' demands. Beyond the classic issues—wages, working conditions, control over the work speed, self-determination of the work pace—a number of new issues have come up in the course of the last three years. I shall recall briefly those issues that seem to have obvious value for industrial workers in other countries and whose politico-cultural importance implies a strategic onslaught against the current workers' condition and the capitalist relations of production.

1. *Unconditional protection of the workers' physical integrity.* Under the slogan "health is not for sale" ("la salute non si paga"), this issue means that it is no longer acceptable that workers, for the sake of capitalist standards of profitability, be subjected to an environment and to working conditions that are damaging to their health.[8] This includes noise, toxic fumes, heat, etc., and also night shifts, which are to be refused wherever they aren't technically inevitable. The capitalist premise that the work force is a commodity among others and that the vital substance of the worker can be bought for a "fair price" (with premiums paid for unhealthy and harmful conditions) is rejected under this principle.

2. *Protection of the workers' cultural integrity.* This issue, which is of much larger socio-political importance, illustrates the fact that *there are no unskilled workers,* and that people who are employed in so-called unskilled jobs are simply denied the chance to unfold their skills or to get recognition for them. This denial is not the result of technological requirements; it originates in technological changes, which are then used as an alibi.

Generally speaking, the lack of skill or loss of skill in blue or white collar jobs is the result of a management policy aiming to eliminate the possibility of worker control over the work process and to maximize control over the workers. In its present form automation reinforces this subordination rather than mitigates it.

As Sergio Garavini[9] and Antonio Lettieri[10] note, the capitalist organization of work tries "to call upon human intelligence as little as possible" and to "hamstring it in the most rigid hierarchical organization," "to the point of mutilating and sterilizing the individual and group faculties" of the workers. "The essential task is to give back to the worker the possibility of unfolding his [or her] abilities and to attain some self-realization through work...by making the best of the potential of scientific and technological development." "We must deny the so-called objectivity of technology and/or the division of labor, expose and denounce its oppressive and exploitative character, and aim at its change by taking the needs of the working person as our point of departure."

In practice the struggle to regain control of work, which is also the struggle against wage disparities, inequality, arbitrary hierarchy, and for class unity, will include the following aspects:

a. The struggle for egalitarian demands (equal raise in pay for all), for recognition that all workers hold some skill (and deserve a corresponding wage), for the elimination of bonuses and their incorporation into the basic wage, for the abolition of job evaluations, and for a single job classification scale with no more than six to eight positions from the lowest to the highest paid member of the personnel. The single scale will aim in particular at suppressing arbitrary distinctions between blue and white collar workers. The struggle for the single scale can only be carried on—and this is its additional intrinsic advantage—when the workers are called to evaluate by themselves, setting their own criteria in free assembly discussions, how many "classes" (or positions on the scale) they think are justified. In the Italian steel industry, the workers agreed on six "classes."[11] This struggle is a first step toward:

b. The abolition of unskilled, repetitive, and stupefying jobs, and the overhaul of the work organization, i.e., job enrichment, job rotation so as to reunify production work, quality control, tool-making, etc., thereby enabling each worker to gain an overall view and potential control over an entire sector of the production cycle. In the long run there should only be two classes of workers, "skilled" and "specialized," so as to make for "a permanent enrichment of the theoretical and practical abilities of each, thereby allowing everyone the full development of his or her capabilities as an individual and as a group."[12]

Steps in this direction are already visible in a few advanced enterprises: it is therefore all the more urgent that the working class take into their own hands initiatives, experiments, and research which are currently carried out by the corporate establishment, so as to impose their own solutions. This would be an important step towards collective self-determination and self-management of the work process and also toward the overturning of the capitalist hierarchy and the cultural mechanisms of bourgeois control.[13]

c. It is impossible not to discuss here the necessary overhaul of the educative system. The general crisis of the school system at all its levels points up the contradiction between the *social* function of the capitalist school and its *educative* function.[14] The grade school pupil or college student doesn't find in school either personal growth or any real instruction. The social function of the school system is essentially discriminatory. It tends to give a cultural basis to social inequality. By inflicting on pupils of all ages the boredom of lessons that are both devoid of intrinsic interest and cut off from life, and by submitting them to a competitive system that bases the success of a few on the failure of others (and their relegation to "inferior" status), the school system handpicks not the most "gifted" but the *most ambitious*.[15] That is, it selects those social climbers who, out of ambition, accept the disciplinary and hierarchical structure of a schooling system whose relations of education prefigure the social relations of production and aim to reproduce them.[16]

The connection between post-secondary school attendance and getting a prestigious job is, however, falling apart. The number of jobs is increasing more slowly than the number of college graduates. The absence of "markets" prevents the graduates of post-secondary education from reaching the promised social position. Except in the extremely selective schools funded by the corporate establishment, post-secondary diplomas are losing their value; the unreality of the "culture" they represent is becoming clear, and the growing mass of students is but a mass of camouflaged unemployed, bound to a new kind of "unproductive forced labor without pay." The "right to study" appears as the "denial of the right to productive work."[17]

Hence the proposals of Italian unionists like Lettieri and Garavini offer a general reduction of the working day (to six or four hours without reduction of pay), which would give all workers the opportunity to study and all students the opportunity to perform productive work.[18] This would be accompanied by *a complete overhaul of the schools* as well as of the organization of work. "Culture" and production, science and technology, intellectual work and manual work would cease to be separate. Schools and factories would stop being ghettos. The social relations of education and the relations of work would undergo radical changes and be collectively self-determined in such a way as to make for the maximum growth of individual and group creativity. The spread of job versatility would make for the abolition of unskilled work and for the constant enrichment and rotation of jobs. It would make concrete the possibility of social and technical self-management, along with the decline of functional hierarchies and of the state.

This proposal also openly aims at unifying the workers, the unemployed, and the students in a situation where the notions of "full employment" and "productive work" tend to lose their meaning, where the right to a living and an income can no longer be made to depend on holding a stable job, and where the rapid reduction in the amount of socially necessary work means that the boundaries between work and culture, between working time and free time, have to be erased.

This enlargement of the union's concerns to include education and culture demonstrates the fact that traditional unionism has become obsolete and that labor has to open up to strata who—like students, unemployed, women, patients, etc.—are its indispensable allies. In particular, labor's offensive against the capitalist organization of work can only succeed if it goes hand in hand with an offensive against a school system that is the cultural womb of social stratification and of the hierarchy of jobs.

The Technical Intelligentsia

Such concerns are even more justified given that the social hierarchization of the "professions" is but a relic that is masking the *de facto* proletarianization of most technical and intellectual workers. In the past, these workers held a monopoly of knowledge and had authority over manual workers. They were part of a privileged middle class to whom some of the bosses' authority was delegated. Thus there was a social and cultural barrier between them and the rest of the proletariat that was tantamount to a class barrier. This barrier still survives in the old fashioned labor-intensive industries where the technicians direct, supervise, organize, set the work speeds, etc., and where they have an antagonistic relationship with the production workers, who are their hierarchical subordinates.

However, in advanced industries, which are partially automated, the technical workers themselves have to perform tasks that are fragmented, rigidly predetermined, and tedious. Even when they supervise and direct the working of automatic operations, they are subject to orders from the machinery, and are bereft of power and initiative.[19] Underemployed, frustrated, and disenfranchised in their professional capacities, these workers can be more sharply aware of their alienation than the manual workers. They are experiencing the kind of *freeze on their professional development* that until now was the common lot of unskilled workers. It had always been the (partially successful) task of the discriminatory school system to convince the unskilled workers that their relegation to the most tedious jobs was the result of their poor performance at school—meaning their "unfitness" to learn and to do better. For the technical workers, though, this was not at all the case. The freeze on their professional advancement and the subordinate positions they occupy seem *arbitrary and unjust*. Management hopes to win their loyalty and devotion to their work—with which it is intrinsically impossible to identify, being unfathomable and without visible results—by offering them symbolic compensations such as status and relatively high wages.[20]

Even though these "incentives" do not really appease this stratum's latent rebellion, it remains a difficult sector to organize and win over to the common struggle on a class basis. To understand this difficulty it is important to grasp the ambiguity of the technical workers' discontent. They are rebelling against their disenfranchisement and frustration not as workers or with the other workers, but as a "separate" group that rejects proletarianization *only regarding itself* and demands the restoration of its old privileges and powers. They tend to protest the hierarchic structure of the workplace and the power and management of the corporate establishment, but their protest is not based on class. Rather, they consider that if they were allowed full use of their abilities they would know how to manage production better and more rationally than the top executives do. In other words, by tradition and training the natural ideology of this group is technocratic and corporatist. It is very far from being the avant-garde of a "new working class."

This doesn't mean that under the right circumstances technical and intellectual workers couldn't become radicalized very quickly, nor that they cannot be won to the class struggle. They can be won to it if, through prompting from their own radicalized avant-garde, they can be brought to see:

• That their proletarianization is an irreversible consequence of monopolist centralization and that their former privileges will never be recovered;

• That they cannot free themselves alone, but only along with the entire working class, by seeking to get rid of the capitalist division of labor, excessive specialization, the separation between jobs of conception and those of execution, and hierarchical structures;

• That beyond the irrationalities they perceive in the way their own workplaces are run there is the much more fundamental irrationality of the capitalist economy. That is, parasitism and waste on the societal scale coexist with the ethic of productivity and efficiency on the scale of each production unit. The overproduction of goods which respond to no felt need coexists with the refusal to fully exploit the liberating potential of science and technology.

The unification of intellectual and technical workers with the working class can only come about insofar as the latter are able to suggest to them that they go beyond their regrets for the past and their corporatist interests, and move toward a wider and more radical view—a view favoring a society in which knowledge would be accessible to all and in which everyone could use it to serve a society of equals and not the firm's selfish crave for growing profits. In this respect, labor would do well in adopting and developing in its own way Ralph Nader's call for a new professional ethic in which scientific and technical workers would put their loyalty to the people above loyalty to their company. In this way they could uncompromisingly fight against all forms of cheating, stealing, waste, environmental destruction, and direct and indirect harm to the physical or mental integrity of human beings as producers and consumers alike.

The Nature and Limits of the Union

A union with a vertical structure and centralized leadership is not capable of organizing and unifying the working class around the issues I have brought up here. A national structure is certainly necessary to interpret in depth the (often latent and badly expressed) aspirations of the workers, to make the politico-ideological implications clear, and to coordinate the struggles. But these struggles will match their goal only if they are themselves the first practical application of the demand for workers' power (or control) and of the self-determination and workers' democracy that this demand implies. Hence the need to "demultiply" the leadership of these struggles by leaving unrestricted possibilities of self-expression, debate, and initiative to the rank and file and its locals. The general issues (themselves formulated by the central leadership after months of nationwide open debate) will be worked out and translated into concrete demands by open assemblies and by revocable action committees which will be free to determine the forms of the struggle and to elect the union representatives. The union will be responsible to these base units and not the other way around.

However, successful democratization and debureaucratization of the union—as was achieved by the Italian metalworkers in 1968-69—immediately brings up the question of the nature and limits of unionism. The independent assemblies give rise to radical methods and militants, to demands that are unacceptable to business and to the government. They give rise to strongly politicized bodies (the councils) with a twofold mandate that implicitly or explicitly brings up issues of political power and the transformation or takeover of state power. The question— brought up at the beginning of this essay—of the transformation of the union into a mass political movement thus takes concrete form. The alternatives for labor are the following:

1. On the one hand it can try to keep the leadership of a movement that is (at least potentially) revolutionary by making itself into a political force among others and working inside the action committees and organs of dual power (such as assemblies and councils) so as to enhance the self-organization of the proletariat, the spread of class confrontations, and the conquest of political power. In this scenario *the union as such would disappear*, and be replaced by the councils in its function of unitarian representation and organization of the class as a whole. This transformation of course could occur only in a revolutionary situation *recognized as such* by a union movement that would be prepared for it.

2. On the other hand, the union leadership may consider that the dynamic of a radicalized struggle threatens to precipitate an economic and political crisis which labor lacks the capacity to bring to a revolutionary outcome. It therefore will do its best to channel the struggle toward negotiable objectives and reformist solutions. In doing this, though it goes beyond the level of classical trade unionism (in its definition of reforms and in its direct intervention into politics), the union leadership will be in conflict with the class avant-garde and will make itself the representative of the "average" masses against them; it will be working toward a negotiated settlement that will be, in its essence, compatible with the survival of the system.

In this scenario (which has always proved the case up to now) the union remains true to its institutional nature. It is a mediating force between the working class and the system. It represents the workers' demands within the framework of the capitalist system and keeps respecting the legitimacy of big business and government. And conversely it represents the logic and the continued existence of the system to the working class. It can only survive *as a recognized institution*[21] and retain its bargaining power if it is capable of translating the workers' demands into negotiable claims and thus containing the class struggle within limits the system can bear. If it fails in this institutional function or gives it up, it ceases to be an acceptable and legitimate partner in the eyes of the ruling class. In this event, class struggle breaks out of its institutional framework to become once more a test of strength and violent confrontation.

Because it is afraid of losing control over worker militancy and of then seeing the struggle escalate to a level that wouldn't lend itself to a negotiated conclusion, the union leadership generally impedes direct democracy and initiative on the part of the rank and file. It opposes assemblies and independent committees, and the election and recallability of all those who bear responsibility. The tensions between the rank and file and the union machine, between the avant-garde and the union leadership, are inherent in the very nature of a union and reveal its ambiguous nature and its limitations.

Therefore, the transformation of the union into a mass political movement can only take place during a period of general confrontation and sharp crisis, when the union's methods are made obsolete by the methods of insurgent masses, and extra-union forces take over the initiative and leadership of the struggle. Outside of a revolutionary situation, it is a mistake on the part of the class avant-garde to consider the union outdated, to call for its destruction, or to try to create a new revolutionary union. The role of the union is not to make the revolution, and the role of revolutionaries is not to make or remake a union. The relationship between union and class avant-garde can only be dialectical and conflictive. In the alternative "negotiated compromise or revolutionary confrontation," which is inherent in most great struggles, the union represents the first choice and the avant-garde the second.

In a great confrontation, the avant-garde are justified in wanting to win hegemony and leadership of the movement—by winning the union militants over to the revolutionary point of view, and by politicizing and radicalizing the struggle much further than the union leadership would accept. But they are in no way justified in trying to win the leadership of *the union* itself. Such an attempt to take over the union machine must be considered absurd. For either the revolutionary avant-garde will prevail, which means the union's structure and logic will be superseded by other structures (the councils); or the struggle will not get that far and, since the confrontation will then have to reach a negotiated outcome, the union structure will prevail and the new extra-union adversary structures will fade away. Because they are antagonistic to capitalist society, these new structures can only be intermittent. They cannot be institutionalized or made permanent unless capitalist society is overthrown.

This brings us to the ambiguousness and the limitations of the idea that *the union as it is* can go beyond unionism. This, as we have seen, can only be true in an at least potentially revolutionary situation. Otherwise, when the union makes itself the champion of this kind of self-transformation while continuing to practice a policy of negotiated settlements, the contradiction between its proclaimed ideology and its practice can mean one of two things:

1. It is trying, by verbal radicalism, to neutralize the influence of radicals on its rank and file; or

2. Conscious of the limited effectiveness of union action and logic, it does not oppose a revolutionary outcome of the struggles, but prepares its militants for this, and opens itself up to something beyond unionism and beyond the capitalist state.

This opening up (this "availability," as the CFDT says) means above all that the union leadership will try to monopolize the leadership of the struggles at all costs, and that it will consider the presence of a critical and restless class avant-garde to be useful leavening, even if its relations with this avant-garde are necessarily discordant during ordinary times.

December 1971

NOTES

NUCLEAR ENERGY

1. See *Le Monde*, 13 and 14 May 1975.

FROM NUCLEAR ELECTRICITY TO ELECTRIC FASCISM

1. Electricité de France is the state owned company producing 90% of France's electric power. It holds a monopoly in distributing power.

2. I think I am the only French journalist to have published quotations from this speech, in *Le Nouvel Observateur*, 1 June 1970.

3. *Le Monde*, 22 November 1974.

4. Taking the example of the Union of Concerned Scientists, a group of physicists working mainly at the nuclear research center of the Collège de France launched an appeal that was made public after four hundred scientists had signed it. The "appeal of the four hundred" gave respectability to the opponents of the nuclear program.

5. Puiseux is one of the top economists of EDF and the only vocal critic of the all-nuclear option within EDF management.

6. Most workers in the atomic industry are affiliated to this union which is highly critical of the French program.

7. Declaration to *Investir*, 24 March 1975.

8. In *Alternatives au Nucléaire*, Presses Universitaires de Grenoble.

9. Their names were withheld in order to protect their career interests. Their study was published by Friends of the Earth (Paris, 1975).

BOUNDLESS IMPERIALISM

1. Much of this chapter is based on an interview with Charles Levinson who was putting the finishing touches on his latest essay *Vodka-Kola*. Quotations of Levinson are drawn from this interview.

2. *Capital, Inflation and Multinationals* (London: George Allen and Unwin, 1971) and *International Trade Unionism* (London: George Allen and Unwin, 1972).

3. Bernard Jaumont, Daniel Lenègre, and Michel Rocard, *Le Marché Commun Contre l'Europe* (Paris: Le Seuil, 1973).

4. See Ernest Mandel, *Late Capitalism* (London: New Left Books, 1975); Bob Rowthorne, "Imperialism: Unity or Rivalry?" in *New Left Review*, no. 69, September-October 1971; Nicos Poulantzas, "L'internationalisation des rapports capitalistes et l'Etat-nation," in *Les Temps Modernes*, February 1973.

LABOR AND THE "QUALITY OF LIFE"

1. This text was drawn up for the "study days" organized by the West German metalworkers union in April 1972, on the topic "The Quality of Life."

2. Barry Commoner, *The Closing Circle* (New York: Knopf, 1971).

3. See "L'apogée du néocapitalisme et ses lendemains," in *Les Temps Modernes*, no. 219, 1964.

4. John Kenneth Galbraith, *The Affluent Society* (Boston: Houghton Mifflin, 1976).

5. Ezra Mishan, *The Costs of Economic Growth* (New York: Praeger, 1967).

6. "Les travailleurs mettent le socialisme à l'order du jour" in *Syndalisme magazine* (CFDT), December 1971.

7. A similar analysis is to be found in the leadership of the Italian metalworkers. According to Bruno Trentin, national secretary of the FIOM-CGIL (the Communist dominated union), "the national network of factory councils is a new force with political power. It foreshadows the obsolescence of the union and of the party and of the separation of the two. The councils must come out of the factory, swarm into the neighborhoods, coordinate their actions and work out their economic, political, and cultural programs" (*Le Nouvel Observateur*, 14 June 1971, p. 37). In the same vein, Pierre Carniti, national secretary of FIM-CISL, says: "To change the condition of alienation, powerlessness, and inferiority of the working class will require not only a change in the political direction of the government or in the balance of forces in Parliament, but a share of the power for the workers, to be won through daily struggle." "The forces of the left" must "question their past performance" and "reappraise their whole way of directing the struggles and politics " (Interview in *Giovane Critica*, no. 28, Rome, 1971).

8. Giovanni Berlinguer, *La Salute nella Fabbrica* (Bari, 1968).

9. "Le nuove strutture democratiche in fabbrica e la politica rivendicativa" in *Problemi del socialismo*, no. 44, 1970. Sergio Garavini is general secretary of the CGIL textile union.

10. "Factory and School," in *Division of Labour, op. cit.* Antonio Lettieri is secretary of the FIOM-CGIL.

11. Lettieri, *op. cit.*

12. *Ibid.*

13. André Gorz, "Technology, Technicians, and Class Struggle," in *Division of Labour, op. cit.*

14. On this subject there is a remarkable convergence among the work carried out independently of each other by authors such as P. Ariès, P. Bourdieu and C. Grignon in France, Ivan Illich at CIDOC (Mexico), and Sam Bowles and Herb Gintis in the U.S.

15. Cf., *Letter to a Schoolmistress* (Mercure de France, 1976), and *Deschooling Society* (New York: Harper & Row, 1971).

16. Cf., Sam Bowles and Herb Gintis, *Schooling in Capitalist America* (New York: Basic Books, 1976).

17. See both Lettieri and Bowles and Gintis, *op. cit.*

18. The same idea is found in the *Theses* of *Il Manifesto* (Paris: Le Seuil, 1972).

19. Cf., Otto Brenner, *Automation, Risiko und Chance*, t. II (Frankfort: EVA, 1965).

20. On this subject see the penetrating analysis of Claus Offe, *Industry and Inequality: The Achievement Principle in Work and Social Status* (New York: St. Martin's Press, 1977).

21. It is true that it was not always an institution. In the "heroic epoch" of its birth the union identified with the working class struggle for the right to organize (and thus with class self-organization). Because this struggle was illegal, it had a radical and subversive character which the institutionalized union has naturally lost.

Chapter IV

Medicine, Health and Society

Introduction

The following essay is a commentary on two books and on part of the literature that inspired them. These two books are *Medical Nemesis* by Ivan Illich and *L'Invasion Pharmaceutique* by Jean-Pierre Dupuy and Serge Karsenty.[1]

My goal is not to draw up a technical indictment against doctors, any more than it is the goal of the authors I am discussing. The technical examples, most often borrowed from American and British studies, are used merely to illustrate the arguments which are the webbing of the essay.

1. Capitalist civilization leads people to consume, on the one hand, that which destroys, and on the other hand, that which repairs the destruction. This fact is the mainspring of the accelerated growth of the past 20 years. But the damage is getting greater and greater and the repairs, in spite of their size and cost, are less and less effective. This is particularly true as regards health.

2. There are more and more doctors and more and more sick people. For the last ten years or so, people in all industrialized countries have been dying younger and are more sickly. This is happening in spite of the expansion of medicine—but also because of it.

3. The most widespread epidemic illnesses—cancer, cardiovascular diseases, rheumatism, etc.—are all degenerative diseases, which are caused by civilization and which medicine can neither prevent nor cure. A growing proportion of the population is struck by these illnesses in spite of the use of more and more elaborate medical technology. Everything suggests that they are linked to our lifestyle and environment. Civilizations that are different from ours are free of them. Of all the factors that maintain health, medicine is one of the least effective.

4. Medicine itself contributes to the spread of diseases in two ways:

a. As a social institution, the duty of medicine is to reduce the symptoms that make the sick unfit for their social roles. By urging people to take illness to the doctor, society keeps them from laying the blame on the fundamental and long-term reasons for their ill health. By treating illnesses as accidental and individual anomalies, medicine masks their structural reason, which are social, economic, and political. It becomes a technique for making us accept the unacceptable.

b. By catering to a mythical idea of perfect health, medicine makes people believe that health can be bought.[2] Every organ, every biological event, every stage of life, every infection, and every death must have its specialist; and health comes to depend on the consumption of drugs and a specialist's care. By thus encouraging medical dependency in the healthy as well as in the sick, medicine lowers the threshhold of illness and adds its own poisons to the industrial way of life.

5. Unlike many animal species, we humans are not perfectly adapted to the natural environment. We can only survive by our labor, that is, by the changes we work on nature. Because of this fact, there is no *state of nature* for humankind; our health and the rules of life it depends on are cultural facts. Far from being the gift of nature, for us health is an endeavor.

6. To be in good health means to be able to cope with illness

as we do with puberty, aging, change, and the anguish of dying...
Overmedicalization prevents or hinders individual coping. It
multiplies the number of sick people. This is what Illich calls
structural iatrogenesis, that is, the structural generation of illness
by institutionalized medicine.

7. This overmedicalization is obviously not the most basic
reason for the continued increase in morbidity in the past ten
years. A more fundamental reason may be sought in the fact that
fragmented wage labor and market relationships destroy the
individual's autonomy and capability to take care of his or her
own life, health, ills, and dying.

8. The basis of health is extra-medical, namely: liking one's
work, environment and community. We are more readily prone
to feel sick when our work and our lives seem external and
tedious. This is also what makes society pathogenic: while
increasing the objective factors of morbidity (cf. the degenerative
diseases), it undermines the existential foundations of health.

9. That is why, from a revolutionary point of view, health
and the problem of health must be demedicalized. Both lie
within the jurisdiction not of the doctor and medicine, but of
hygiene.

Medicine, in fact, is the system of codified care and treatment
dispensed to people by a body of specialized professionals.

Hygiene is the comprehensive set of rules that people observe
by themselves to maintain or recover their health.

When medical knowledge becomes part of popular culture, it
brings with it hygienic practices such as washing one's hands,
purifying drinking water, a varied diet, exercising, etc. And these
are what make it the most effective. The difference between
hygiene and medicine is the same as between popular culture and
high culture.

10. Translating useful medical knowledge into hygiene is a
traditional revolutionary goal. It doesn't come from an anti-
scientific attitude, but from an anti-elitist attitude. According to
Illich, nine-tenths of all effective medical knowledge consists in
simple, inexpensive treatments that is within the capability of any
motivated lay person who can read directions. However, the bulk
of medical expenditures goes to hard and expensive treatments
whose effectiveness is unproven.

The goal of this essay is not to get people to refuse all medication and medical care, but to "recover power over their illness, their body, and their spirit. Let them blame the things in their daily lives that make them sick: the school, the factory, the mortgaged suburban house, the couple, etc."[3]

1. Medicine and Illness

For about ten years, medicine has been making more people sick than it has been curing. It has become, of all industries, the most wasteful, polluting, and pathogenic. By claiming to patch up case by case those populations that are becoming more and more sickly, medicine masks the deeper causes of their diseases—which are social, economic, and cultural. While claiming to relieve all suffering and distress, it forgets that in the final analysis people are damaged in body and soul by our way of life. Medicine, in helping them to put up with what is destroying them, ends up contributing to this damage.

Briefly summarized, these are the central theses of Ivan Illich's book, *Medical Nemesis*. It will shock people even more than his previous essays. For medicine, more than speed, than school, than the mega-tools of mega-industry—the targets of Illich's previous work—is a sacred cow. Of all the instruments of our social normalization and our alienation from ourselves, medicine—which attacks the deepest source of our possible personal autonomy, our relationship to the body, to life, to death—is the one enjoying the highest prestige.

Whether professional or lay, don't we attribute to medicine the rapid rise in life expectancy: 20 years at the time of Christ, 29 years in 1750, 45 in 1900, 70 today? Don't we generally attribute to Pasteur and Koch, to vaccinations, chemotherapy, and antibiotics, the drop in infectious diseases and the increase in longevity? Don't we take it for granted that a population's state of health depends on the number of doctors and hospital beds available to it, the amount of medical care and medications they consume? Well, that is all wrong. The efficiency of curative medicine is and always has been limited. It is time to put it back in its place.

The False Triumphs

Indeed, medicine has learned to treat almost all the infectious diseases, some endocrinary deficiencies such as myxedema, and some metabolic disturbances such as diabetes. But on the whole, it is still waging a battle long overdue. For it is powerless to cure the great chronic degenerative diseases that have taken over from the infectious ones as the principal cause of premature death.[4] It remains powerless against the most widespread ills: rheumatism, migraines, diseases of the respiratory tract, and digestive troubles. And, on further reflection, it isn't even true that it played a decisive role in the drop in infectious diseases—for which it is commonly given credit.

Look at the curve below taken from a study by Winkelstein and French.[5] It shows the death rate from tuberculosis in the United States since 1900. What does it teach us? This: in America, as in Europe, tuberculosis killed 700 out of 100,000 people a year at the beginning of the 19th century. In 1882, the year Koch discovered the bacillus, tuberculosis had already declined by half. In 1910, at the time when the first sanatoriums

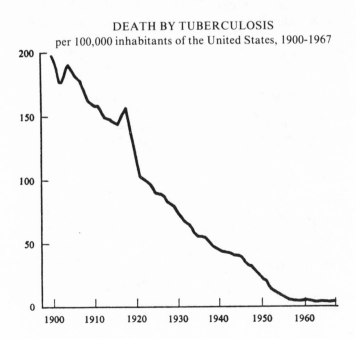

DEATH BY TUBERCULOSIS
per 100,000 inhabitants of the United States, 1900-1967

were built, tuberculosis had already declined by three-quarters. And following this, neither chemotherapy, adopted around 1945, nor antibiotics, used successfully around 1950, have had any real effect on the slope of the curve.

In short, the drop in tuberculosis is not due to curative medicine. Even with equal medical care and supervision, the poor are still four times more likely to get it. Indeed, medicine has perfected more and more effective treatments, but essentially the battle was won without it.

The same kind of demonstration can be made for other great scourges, cholera and typhoid, for example, which today any nurse and even any lay person can treat easily and effectively. For typhoid and cholera had practically disappeared from Europe even before the bacillus and the vibrio that cause them were isolated.

Again, look at the following diagram, for which we are indebted to R.R. Parker.[6] It shows that in Great Britain death from scarlet fever, diphtheria, whooping cough, and measles had already declined by 90% by the time compulsory vaccination and antibiotics were introduced. The slope of the curve has not changed noticeably since these were introduced in 1948.

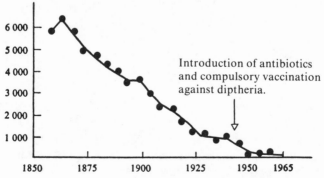

NUMBER OF DEATHS PER 1 MILLION CHILDREN
(less than 15 years old)
Attributed to Scarlet Fever, Diptheria, Measles, and
Whooping Cough (England and Wales, 1850-1965)

Introduction of antibiotics
and compulsory vaccination
against diptheria.

Thus, the infectious diseases were declining independently of the (nevertheless effective) weapons that medicine brought to bear against them, to be replaced by other epidemic diseases against which medicine can't do much. "Industrialization," writes John Cassel, "was accompanied at its beginnings by the rapid advance of tuberculosis. This disease reached its peak in 50 to 75 years (around 1800-1825), then declined steadily, independently of treatment, and was replaced by malnutrition syndromes such as rickets (in Great Britain) and pellagra (in the United States). For partly unknown reasons, these diseases in turn declined and were replaced by the childhood diseases. The rapid decline of these during the 1930s went hand in hand with the spectacular rise of duodenal ulcers, mainly in young men. This affliction in turn declined for completely unknown reasons, to be replaced by the modern plagues: cardiovascular diseases, hypertension, cancer, arthritis, diabetes, psychological troubles."[7]

In the final analysis, diseases appear and disappear in some relation to factors such as the environment, living conditions, lifestyle, and hygiene. Thus the disappearance of cholera and typhoid, the quasi-disappearance of tuberculosis, malaria, and "puerperal fever" are due, not to therapeutic progress, but to the treatment of drinking water, the spread of sewers, better working, living, and eating conditions, the draining of swamps, and the use of soap, sterile cotton, and scissors by midwives and obstetricians. Doctors contributed to the development of these preventive practices, but they didn't become effective until hygiene and antisepsis (like contraception in other situations) stopped being medical techniques and became common behavior. Hygiene, not medicine, guarantees health—hygiene (*hygieia*), in the original sense of all the rules and circumstances of life.

"Even in almost all the underdeveloped countries," writes Charles Stewart, "improvement in the general state of health was obtained almost entirely by the improvement of public hygiene. The increased availability of medical care played only a marginal role, if it played any at all... The fact that for the past two decades life expectancy has hardly advanced at all in the United States and that it is higher in several countries where medicine is at a much lower level than ours suggests that the productive capacity of our health care system is very low."[8]

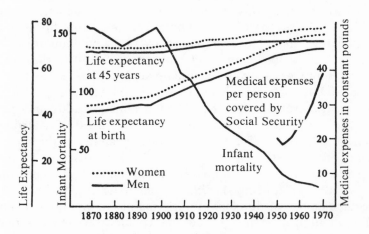

The preceding graph, published by John Powles, is an arresting graphic presentation of this "low productive capacity."[9] It puts side by side the increase in health care expenditures, on the one hand, with life expectancy at birth, on the other. It shows that the doubling of health expenditures since 1950 has had practically no effect on longevity and that the rise in life expectancy since 1920 can be attributed almost entirely to the lowering of the infant mortality rate.

The table presented by Charles Stewart (p. 192) is no less eloquent.[10] It shows almost identical life expectancies in countries that are very unequally "developed" from the medical point of view (the density of doctors varies by a ratio of one to four).

Stewart reasonably concludes that if people die, it is not the fault of the doctors—nor is it if they live to an old age. Would you argue that a person could live to an advanced age without doctors in Jamaica but not in Canada or in France? This argument would say the richer a country is the more subject are its people to diseases. Thus it needs a huge amount of health care in order to live in as good health as a poor country. As pleasing as this may be to ecologists (and to doctors), this argument doesn't hold. No one has ever been able to prove that the increase in the ratio of doctors to population and expenditures for health care have lowered the rate of illness in any one country. The contrary may even be true; this we will see in what follows. If people die younger in some regions in France where there are fewer doctors to care for the population, the reason is very simply...alcoholism. When the effects of alcoholism are factored in, the disparities disappear.[11]

On the other hand, according to Stewart's calculations, there are two factors that do have a very strong positive effect on life expectancy: the piping in of drinking water and literacy. By themselves these two factors explain 85.5% of the disparities in life expectancies around the world.[12] In view of these facts one cannot help but ask what, then, accounts for the astonishing expansion (10 to 15% a year in constant money) of "health" expenditures in all the industrialized nations. What is the meaning of the public outcry for more doctors, more hospital beds, and more drugs? If Americans, who spend $320 a year per person on medical care, are not any better off than Jamaicans, who spend $9.60, why do they waste their money? And why level the expensive (and not very successful) attack at curing the diseases, instead of at eradicating their causes?

One key to these questions may be found in the following fact: more than three-quarters of all health care expenditures in the rich nations are aimed not at taking care of disease, but at taking care of a health that is thought or feared to be in danger. The goal is no longer to restore health, but to preserve it and improve it. And since there is no limit to improvements, an inexhaustible market is available to the makers of "preservatives," stimulants, rejuvenants, fortifiers, tranquilizers, etc. This is what Illich calls "the medicalization of health." (We shall return to this.)

The adage that says "better is the enemy of good" applies here better than anywhere else. Better health is the enemy of simple good health. In suggesting to everyone that we are in danger of a disease against which we could perhaps protect ourselves, by examinations, preventive treatment, and continual care, medicine creates sick people—its own sick people. And this is not at all hypochondria.

Pathogenic Medicine
  a. Direct Iatrogenesis
    There are two kinds of medicine-induced illnesses, those due

to physical intervention by doctors (poisoning, infection, mutilation, wounds, etc.) and those that doctors bring on or maintain by getting people who were not conscious of any illness to adopt the behavior of the sick—to become anxious, self-observant, fearful of strain, dependent.... In the United States there is extensive literature on these "iatrogenic" illnesses and neuroses. Illich adds to these two categories a third: "structural or existential iatrogenesis." By this he is suggesting that medical and pharmaceutical encroachment—the medicalization of health, illness, pregnancy, birth, sexuality, and death—have ruined the ultimate basis of our health: our capacity to take charge of our own physical state and to face by ourselves the events and the trials of our biological existence.

But let us start with iatrogenic illness in the narrow sense.

"The pathogenic effects of medicine," writes Illich, "are of all plagues the one that spreads the most rapidly. Diseases brought on by doctors are a greater cause of increased morbidity than traffic accidents or war-related activities." Exaggeration? Judge from these examples. (They are mostly drawn from U.S. studies, not because hospitals are worse in the United States, but only because the questioning and examination of the health system are more open and critical there.)

In 1965, seeking to measure the risks to which hospitalization exposes patients, whether sick or not, two doctors worked out the following statistics.[13] Twenty percent of all patients admitted to the hospital for treatment or examination were victims of one or several incidents. There was, on the average, one incident per 41 patient-days, one serious incident per 99 patient days. Examination or treatment was the source of 27% of the incidents; 28% of them were due to accidents or mistakes (notably in the administration of medication), and 45% to reactions to the medication. This last cause then obliged 9% of the patients to prolong their stay in the hospital.

This small-scale and local inquiry was of course disputed. The National Institutes of Health (NIH) organized a national inquiry. Its statistics were even more distressing. Out of 32 million people who had contact with hospital care in 1970 (a figure that includes outpatient consultations), more than 10% had to be kept longer than anticipated because of the medication they received.

Furthermore, 1.5 million people were hospitalized following troubles brought on by medication prescribed by "their" doctors.[14] Shortly afterwards a pharmacist, Marc Laventurier, and a doctor, Robert Talley, estimated that at least 30,000 people die every year in U.S. hospitals of medication poisoning. Disputing this estimate, the pharmacists' association and the AMA held their own inquiry, choosing the University of Florida hospital for their investigation. The result caused consternation. In this model hospital one patient out of 555 (about 0.18%) is killed every year by the way medication is administered. Shortly after this, a "medicational mortality" of more than twice that, or 0.44% a year (which is about one patient in 288), was established for Boston hospitals, which have an especially high reputation. (This is the same rate found in Israeli hospitals.)

In short, medications in hospitals alone kill between 60,000 and 140,000 Americans a year and make 3.5 million others more or less seriously ill.[15]

How many patients sustain injuries that are not drug induced? Out of 6000 people in France who die annually "on the operating table" (of which 2000 are the anaesthesiologist's fault[16]), how many are being operated on without any real necessity? How many suffer their whole lives because a surgeon operated on them for a kidney ptosis which did not cause them any suffering at all? How many women have their reproductive organs removed (hysterectomy) unnecessarily?

Another American doctor set out to answer this last question. He arranged to have sent to him the reports on 6284 hysterectomies done in one year in the 35 private hospitals in Los Angeles.[17] He found that 5557 (88.5%) were performed without their necessity having been previously established. According to surgeons' own reports, they did not discover anything wrong in 819 patients (who thus had their healthy organs removed). Nearly half of those operated on (48.2%) had had no other symptom than a "backache." Some had no symptoms at all (5.4%). Worse: after it was too late, 30% of the young women (between 20 and 29 years old) operated on turned out to be free of disease. Post-operative diagnosis failed to justify the operation in all but 2494 cases (40%). In short, they "take it out" and examine afterwards. And that's not a peculiarity of California.

Illich sums up the situation when he says 'accidents are the main cause of infant mortality; hospitals are the places where the most accidents occur. Furthermore, the accident rate is higher in hospitals than in all other industries except construction and mining.... University hospitals are the most pathogenic of all. One in five patients contracts an iatrogenic disease, which usually requires special treatment, and leads to death in one case out of thirty. With an accident rate like that on his record a military officer would quickly be relieved of his command; a restaurant or a night club would be closed by the police."

b. Medicalized Health

What can we conclude from this? That we need more modern hospitals, more and better trained doctors and medical personnel, stricter controls, and higher standards? Illich draws the opposite conclusion. He says medicine has become a hypertrophied industry; its factories, its bureaucracies, its bosses, engineers, and foremen have gotten hold of everything connected to health and illness, expropriating both of them. People are encouraged to go to "those who know." Healing, physical and psychological equilibrium, are no longer thought to be the result of "the art of living," "virtue," and "hygiene" (*hygieia*) in the classical sense, but of constant technical intervention. Those in charge of this intervention have persuaded people that in order to live, survive, get well, or bear their illnesses, they need to live inside a kind of therapeutic bubble in which they are drugged, antisepticized, tranquilized, stimulated, regulated, and permanently controlled.

Medicine has been able to make everyone dependent upon it, because this fundamentally pathogenic society has indeed produced a fundamentally sickly population. Health professionals, far from attacking the deeper causes of illness, limit themselves to keeping track of it and isolating the symptoms. They only offer to reduce discomfort, to mask pain, to relieve people of their anguish, to keep things from getting worse. Medicine thus becomes the technical ritual of a type of control which in fact makes use of incantation and magic (renamed "suggestion," "placebo therapy," "securitization," etc.) and saps personal autonomy more radically than the priests ever did.

"Medicalized" people no longer consider it natural to fall ill and get better, to grow old, and to die. "In our day," says Illich, "no one is ever carried off by death, but by a disease from which one 'should have been able' to be 'saved.' You don't recover from an illness anymore, you are cured." No one is well, but rather well cared for, well protected against an infinity of troubles, for whose signs people must always be on the look out.

It is more through the medicalization of health than through the medicalization of illness that medicine ends up making people sick who, without it, would consider themselves well. To say that it makes more people sick than it heals is not a rhetorical exaggeration. To object that the risks medicine exposes us to are small compared to those of the illnesses that threaten us ignores this primary fact: 90% of the time people get well (or can get well) without therapeutic intervention. According to the NIH report cited above, 60% of all medication and 80 to 90% of all antibiotics are administered needlessly.

But here is a second fact, which the press made a big deal of at the time. During a hospital strike in Israel (which lasted a month), the Israeli mortality rate was lower than at any other time. Only emergencies were admitted, which lowered the usual number of admissions by 85%. This same reduction of 85% was recorded during the strike in the New York City hospitals. It was as if the population were in better health when medical care was limited to emergencies.

c. The Early Detection Trap

But, you may say, what about preventive medicine? Doesn't it reduce the risk of illness? Well, as we shall see, quite the contrary. When, as is now the case, medicine that is called preventive doesn't bother about making working and living conditions healthy but only about tracking down early signs of diseases, it increases the number of sick people instead of reducing it. First of all, as I. Boltanski noted, compulsory check ups "lower the threshhold of tolerance to sick feelings, build a more self-indulgent relationship to the body, increase insecurity," and because of this, "increase the subjective possibility of illness and medical recourse."[18]

But there is still more. Some Americans have tried to measure how pathogenesis through early detection operates. For example, here is a study by Bergmann and Stamm on early detection of cardiac disease in schoolchildren.[19] Amazed at the number of children who were forbidden to play sports and games, protected at school, hovered over at home, and stuffed with sedatives because they had a "heart murmur," Bergmann and Stamm studied the entire school population of Nashville. Their conclusion: 44.4% of the children had a mild "heart murmur" that didn't keep them from feeling perfectly well. "The disabilities inflicted on them because of this non-disease seem more serious than those that would be brought on by the disease, if it existed."

Audy and Dunn, in another study, looked at the following train of events. They examined 4000 people who were feeling well and confirmed that 30% were clearly ill without being aware of it, and that 60% had latent diseases to which they were well adjusted. Only 10% were in clinically good health.[20] The authors' conclusion: when these people who were feeling fine were informed of their clinical profile, that was all it took to transform 90% of them into patients and bring on in most of them the appearance or worsening of symptoms that they had ignored up to then.

Do you argue that by treating them at the presymptomatic stage there is a better chance of halting or curing the detected diseases? Well, disabuse yourself. Anxiety incited by a diagnosis or prognosis causes a general deterioration of health. N.J. Roberts verified this in a study that included several thousand patients followed over seven years. Treatment of diseases from the presymptomatic stage is only half as effective as treatment after the symptoms are already manifest.[21]

Add to this that laboratory tests are often wrong or wrongly interpreted, even when they involve simple measurements. In hospitals, where blood tests and measurements of glucose and urea are regularly done, 67% of all abnormal results go unnoticed. "Doctors are so flooded with normal results that abnormal results escape their attention."[22]

Without getting into the vaccination argument, here is nevertheless a recent piece of news. Twenty years ago, before the introduction of the compulsory vaccination, 100,000 British children caught whooping cough every year. Around 160 died of it. Currently, according to Professor George Dick of Middlesex Hospital Medical School, 80 children a year die as a result of their vaccination and 80 others incur irreparable brain damage.[23]

But, people will ask, isn't it better to detect cardiovascular diseases and cancers, which are the apparent cause of 66.7% of all deaths, as early as possible? Well, suppose after a checkup you are assured "everything's fine." That doesn't protect you against a heart attack the following week or month. On the contrary, remarks Paul Clote, "a reassuring bill of health can encourage a patient to ignore symptoms that may come on shortly afterwards," while with no bill of health at all, the person would probably have been careful not to overwork him or herself.[24]

Assume on the other hand that the checkup confirms what you were worried about and what anyone could have told you without using complicated and costly technical devices. You have high blood pressure, a "weak heart," you should stop smoking, eat less, exercise more, and take more time off. In short, change your way of life and your social and professional ambitions. However, the typical heart attack candidate is precisely someone who, rather than lowering ambitions, accepts the risk of being "struck down in mid-effort." The hygiene that could save the individual from heart attack would be a professional handicap. Medicine cannot help.

Thus, as Clote writes, "early detection of a cardiovascular problem is of very little use since there is no positive way of diminishing or stopping the disease." There are certainly pills to bring down blood pressure, but their secondary effects are considerable and there is no evidence that their advantage is greater than their risk. As for the attempts at *medical* prevention of cardiovascular diseases, they have persistently failed. Experiments with preventive medications, tried in the United States, were abandoned at the end of 19 months because the group taking the medications "had a higher death rate and a greater number of pernicious effects (infarcts, embolisms) than the control group receiving the placebo."

In short, what good is it to detect diseases that medicine can neither treat nor cure? This is also a question to be asked about most cancers. Is it absolutely necessary to detect a lung cancer when 95% of the patients who are operated on following an early diagnosis still die within five years, so that the main effect of the diagnosis is to spoil the little time they had left to live "normally"?[25] Is it absolutely necessary to "treat" breast cancer when—beyond an early stage during which it is curable—70 to 80% of all women treated will still die within six months to two years of secondary cancer, and the months left for them to live will be ruined by treatments (radiotherapy, mastectomy, chemotherapy), all of which are very trying?[26]

Concerning the cancers that lead to a death that is often painless, Turnbull notes that "the surgical or radiological treatment, when it eradicates the primary disease, allows the development of a secondary cancer which itself is often painful.... The price of healing is often greater than we admit."[27]

This is also Illich's point. Along with Paul Clote, he maintains that "the only effect of early treatment of incurable diseases is to make the condition of the patient worse," whereas in the absence of all diagnosis and treatment "the patient would stay feeling well for two-thirds of the time left to live." If these statements are shocking, that is only because it has become shocking to state that *it is natural to die*, that there are and *always will be fatal diseases*, that they are not an accidental and avoidable irregularity, but the *accidental form taken by the inevitability of death*. And when everything is taken into account, it might be better to die of the disease one has rather than the iatrogenic or secondary one contracted in its stead.

Healthy good sense has become a scarce item in our medicalized civilization. Medicine's very recent contention (it began in 1920) that all illness must be or become curable has transformed any sick person's death into an "accidental death," and given birth to the idea and the ideal of the "natural death." This is a death without pathological cause which comes about because the body is worn out, and, used up but intact, it goes out like an oil lamp. The ideal of the "natural death" is to die in good health after having marshalled all the resources of medical technology.

Thus, we have medicalized death along with illness, health, and birth. Whoever doesn't die in the hospital dies an irregular death which must be the object of a legal or medico-legal examination. To abide by the rules, you must die in the hospital. And at the hospital, of course, you can only die with the doctors' permission. Your death, like all your illnesses and your health, becomes the business of professionals. It doesn't belong to you. The art of dying (*ars moriendi*)—which implied the farewell ceremony in which the dying person, surrounded and assisted by all his or her near ones, sums up the meaning of his or her life and "dies as he or she lived"—has been replaced in our culture by the clinical death, solitary, shameful, and absurd.[28]

The circle is now complete. The modern man or woman is born in the hospital, taken care of in the hospital when he or she is sick, examined at the hospital to see if he or she is feeling well, and sent to the hospital to die by the rules. We are robbed of the last foundations of our autonomy by the same mega-machines and mega-institutions (born of the concentration of capital and the spread of market relations) that rule the rest of our life. Illich: "The person who learned by seeing and doing, who moved on his or her own, who had children and brought them up, who got well and took care of his or her health and the health of others, has given way to the person riding a motorcar, giving birth in a hospital room, being educated at school, and cared for by health professionals." For every need we have becomes dependent upon commercial goods and services, dispensed by institutional machinery that is beyond our control and which breeds dependency, scarcity, and frustration:

• Increased vehicular speed paralyzes our transport system and causes us to lose more time in transit than ever before in history.

• Chemical agriculture is destroying fundamental ecological balances and is putting the world on the threshhold of further famines.

• Schooling makes us incapable of learning by ourselves and destroys our very desire to do so.

• The spread of wage labor and of large scale commercial production makes us incapable of producing according to our needs, of consuming according to our desires, or of defining and leading the life we want.

• Finally, the medico-pharmaceutical invasion makes us more and more sickly and destroys the deepest springs of health.

All of this is what Illich calls *Industrial Nemesis*, of which *Medical Nemesis* is only one aspect. We have traded off enslavement to nature for enslavement to an even more terrible and tyrannical antinature. And we have lost health in the process.

For health is not a biological given. It is, says Illich, "a task: the capacity to adapt to a changing environment (within certain limits, obviously), to grow, to age, to get well when stricken ill, to give birth, to suffer, to face death calmly...to live with one's anguish.... When the need for specialized care goes beyond a certain point, we can gather that society's organization and goals have become unhealthy.... To stay in good health then becomes a task requiring the subversion of the social order."

Here, then, is the heart of the problem: in a pathogenic society health is also a political task. Just as a specialized institution, such as the school, cannot truly educate when social life has ceased to be educational, medicine cannot confer health when the way of life and the environment are injurious to it. Anthropologists and epidemiologists know this well. People do not only fall ill from external and accidental causes which can be cured by means of technological care. They also, and more often, become sick from their social and personal lives. Medicine which claims to treat diseases without considering their sociogenesis can have only a very equivocal social function. At best it can be an act of charity in which the doctor takes over the empty place of the priest. At worst, it is an industry that encourages sick people to continue their unhealthy way of life for the greater profit of manufacturers of all kinds of poisons.

But, rather than judging, we must ask: Why is medicine what it is? Why has the public everywhere such an insatiable appetite for it? Are "demedicalization" of health and "deprofessionalization" of health care imaginable?

2. Health and Society

In nine cases out of ten, there is no point having a medical professional diagnose and treat a common illness. The symptoms are clear, the remedies well-known and very cheap, and, if they promote healing, these medical professionals are not necessary for healing. Also, in China it only takes three weeks to train a "barefoot doctor," who, while continuing to work as a factory or farm laborer, will know how to treat common afflictions, dispense medication (for which he or she is perfectly able to recognize the counterindications and incompatibilities), and recognize the cases that require a specialist; and all this with an accuracy that arouses the admiration of Western doctors who have been on the spot.[29]

According to a Canadian report, which is cited by Ivan Illich, the cost of curative medicine is so low that if the present health expenditures of India were equitably allotted, all Indians would benefit from it.

According to the director of the World Health Organization, the diagnosis and treatment of skin diseases can be learned in a week by anyone with a college degree.

According to a Chilean medical commission comprising Salvador Allende (who was himself a doctor), for all diseases there are only a few dozen medications that have demonstrable therapeutic value; consequently pharmacopeia could easily be scaled down. More than half of these medications could be freely sold over the counter accompanied by direction for use.[30]

And yet for the past 20 years, in all the industrialized countries, medical apparatus and expenditures labeled "health" have increased wildly, at two or three times the speed of national production. From 1950 to 1970, per capita "health" expenditures (in constant money) grew by a factor of 3.5 in the United States, of 4.6 in France, of 2.1 in Great Britain (where the increase was the slowest).

The growth of pharmaceutical consumption has been even more rapid. In France, per capita purchase of drugs has increased, in constant money, by a factor of 2.7 in 13 years (1959-1972). According to a British pilot study, more than half of all adults and almost a third of all children take some medication every day. In Great Britain and the United States, there are as many renewable prescriptions for psychotropic drugs (tranquilizers, sleeping pills, etc.) as there are inhabitants of the country. The U.S. pharmaceutical industry produces 18 doses of amphetamines and 50 doses of barbiturates per person a year.

Even so, this orgy of medication and professional attention is ineffectual when it comes to the improvement or prolongation of life. On the contrary: in France life expectancy of people over 60 is only two years higher than it was in 1900. For French men in general it hasn't risen since 1965. In the past ten years or so the mortality rate of men in their forties and fifties has risen in all the industrialized countries. The mortality rate for young people between 15 and 20 years old is rising by 2% a year in France. For British workers over 50 it is now higher than it was during the 1930s.

Would you argue that the mortality rate is not necessarily a good indicator of general health? J.N. Morris thought of this objection while refining his statistical investigations. In doing so, he established that the deterioration of the general health (or the growth in morbidity) was worse than the increase in the mortality rate had led him to fear. In 20 years "an appreciable increase of chronic diseases in men from 55 to 60 has occurred, and in men entering their sixties the increase is higher by something like 30%."[31] The British National Health Service notes in its 1970 report that in a six year period (1963-1969), the number of days lost because of illness has increased by 20%. The increase was particularly high for cardiovascular diseases, rheumatic problems, and respiratory tract diseases other than bronchitis and tuberculosis.

These statistics give the lie to the common statement that "if there are more sick people, it's because people are living longer." They also contradict just as clearly any belief in the ability of increased health care consumption to improve the general health. The truth is much simpler. *People are medicating themselves more because they are more morbid, and the very rapid increase in their medical consumption doesn't at all keep their morbidity from increasing right along with it.*

Medicine thus appears to be poorly suited to the goals it claims to pursue. Its development no longer brings any benefits, and ends up causing more damage than it repairs.

How can this be explained? Essentially, by the fact that our way of life and our environment are becoming more and more pathogenic. Degenerative diseases, like the infectious diseases whose place they have taken, are basically diseases of civilization. Rather than calling them by the name of the part of the body they affect, Winkelstein says, we ought to name them and classify them by their causes: diseases of affluence (due to overeating, sedentariness, tobacco, etc.), diseases of speed, diseases of modern conveniences (due to lack of exercise and natural foods), diseases of pollution, etc.

Recent studies have established that cardiovascular diseases, hypertension, and, in particular, hypercholesterol are very rare in so-called primitive people, no matter what their age. They afflict the aging in our civilizations alone.

Furthermore, cancer of the colon or rectum, which is the tenth most common disease in men, is ten times more common in industrialized countries than in rural areas in Africa. It seems to be encouraged by a diet whose lack of bulk (that is, indigestible fibers) seriously slows down intestinal transit.

Dr. Higginson, of the International Agency for Cancer Research, estimates that 80% of all cancers are caused by the way of life and the environment of industrial societies. Stomach cancer, for example, seems to be linked to air pollution from coal smoke.[32]

Respiratory tract and lung cancer are linked to the inhalation of tobacco smoke. According to the British cancer specialist and epidemiologist R. Doll, "many indications lead us to believe that most cancers are caused by the environment—notably the fact that incidence of cancer varies widely from country to country and that this variation is reconfirmed in groups that migrate from one country to another. It follows that most cancers could, in theory, be avoided."[33]

Other British and American statistics reveal that the mortality rate from lung cancer and chronic bronchitis is twice as high in cities as in the country. Lare and Saskin estimate that merely lowering air pollution by half would reduce death from lung cancer by 25%, from bronchitis by 50%, from cardiovascular diseases by 20%, etc.[34] According to Eli Ginzberg, "a diversified, fiber-rich diet would contribute more to a population's health than any new medical developments."

But these truths are still ignored or without influence. It is as if medicine, doctors, health policies, and the public prefer to care for the sick rather than to prevent diseases. The health of the healthy seems to have become so stripped of value that it is stupidly damaged almost as an institutional policy by industry, public agencies, and people themselves. On the other hand, "life has no price" when it's a matter of "saving" a small minority of sick people or "repairing" damages with the big and very expensive machinery of advanced medicine.

So it is not surprising that the cost of medicine is growing by leaps and bounds (particularly that of hospital care) even while its returns decrease. How could it be otherwise when medicine neglects the most effective measures (which are preventive) and invests in spectacular performances whose effectiveness is doubtful and whose cost is so high that most people can never benefit from them?

For example, look at the technology of organ transplants: whatever its scientific repercussions, there will never be enough organs to transplant into all who need them. Neither will there ever be enough life support services for all the dying whose lives—and suffering—have some chance of being prolonged.

Look at intensive care units, which are veritable factories of advanced medicine, intended to save heart attack victims. Like all life support services, these units need three times more equipment and five times more specialized personnel than an ordinary care unit. Is it nevertheless necessary to create hundreds of these—plus the helicopter networks necessary to make them available to rural people—without regard for the expense?

A British investigative commission, headed by Lord Platt, studied the issue. It concluded that intensive care units had no demonstrable advantage over care at home. Furthermore, the commission said, "more than half of all deaths occur before the doctor arrives, and most of the lost time elapses even before the doctor is called. We can thus say that 50% of all heart attacks that lead to death are beyond all possibility of medical treatment. For these cases we have to bet on prevention."[35]

But real prevention has no appeal. As Jean-Pierre Dupuy has very well demonstrated, it pays off politically to install a new supermodern hospital unit, but a politician could expect little gratitude from the voters were the number of sick people reduced by half.[36] The people whom preventive measures keep from falling ill have but a statistical existence. They are "statistical persons." Unknown to all and to themselves, they are not grateful for the protection from which they have benefited. Who would ever say, "I'll vote for Deputy Mayor So-and-So because thanks to him (or her) I haven't been sick all year"? In contrast, the sick person who is taken to the hospital is a specific person who, along with all his or her family, will hear the Deputy Mayor when he declares, "I am responsible for this new hospital; vote for me."

But it is not only in the political sense that illnesses, which are cared for rather than prevented, "pay off." Illness turns the wheels of some of the most profitable industries, creating employment and thus "wealth." The simultaneous growth of the number of sick people and "health" industries appears on the plus side in the national accounts, whereas the disappearance of these industries, should they lack sick people, would translate into a reduction of the GNP and would be a hard blow to capitalism. In short, illness is profitable, health is not.

This is why medicine continues to develop in a way that goes against all good sense and fair play. In the same way that more importance is attached to the performance of the Concorde than to the daily transportation of millions of suburbanites, there is more interest in the adventurous pioneers of advanced medicine than in the preservation of health. The result is that the development of medical techniques (as with transportation) creates more want, inequality, and frustration, and consequently satisfies fewer needs. And all along it maintains the worst of illusions: that medicine will soon know how to cure all diseases and therefore it isn't important to prevent them.

This illusion is found even in the medical vocabulary itself. Aren't checkups and early diagnosis of degenerative diseases called "prevention," even though there is no treatment or remedy for them? John Cassel puts it very well: "We have never been able to prevent diseases by detecting them in individual sick people, but by dealing on a collective level with the environment and with the social and psychological factors that increase vulnerability to a disease and weaken resistance. Health is essentially a balance between pathogenic agents and their hosts. It depends on the individual's capacity to maintain a relatively stable relationship with his [or her] environment.... *The important point is to learn how this capacity can be given social support.*"[37]

Increasing morbidity, indifference to true prevention, spectacular overconsumption of health care, and medications that do not reestablish health: how do medicine and doctors accomodate themselves to this absurd situation? To blame them is too easy.

Their conceptions of the sick person, of illness, of the function of medicine, are still profoundly marked by the bourgeois ideologies of the 18th and 19th centuries. The body is seen as a mechanism whose cogs are out of order. The doctor is the engineer who puts them in working order again by surgical, chemical, or electrical means.

Unlike ancient medicine, bourgeois medicine knows only individuals, not populations. This is appropriate, of course, to the relationship the doctors have with "their" patients. They are private individuals, customers, and they ask that the doctors relieve their pain, cure them, advise them—here and now, as they are, in the world as it is. The doctors conform to this demand. That is their trade. No one asks the doctor to see beyond individual cases to the social, economic, and ecological causes of the disease. In this way medicine is turning into a bizarre "science" that studies partial structures minutely without taking into consideration the whole structure to which they belong.

Only a few pioneers, missionaries, and crazies are interested in the epidemiology and the biology of whole populations, or in anthropology, or in work-related diseases. These true researchers and theoreticians, while they preserve the honor of the medical profession, have no influence on the practice and function of medicine. No money is available for studying the health of *populations*, no one pays doctors to bother about it; moreover, their training and social position do not prepare them to advise people how to make their habits and their environment health supporting.

Thus they practice their profession within the narrow limits of the social system, conforming to the social norms in an astounding way. How, wonders Powles, could they not have predicted that inhalation of gases and chemical vapors, of smoke (tobacco, molten metals, heated oil, coal), and dust (asbestos, cotton, granite) would be extremely injurious to health? How could they not have risen up against the conditions of life in industrial and mining towns, conditions whose ravages they saw every day? Don't they refuse to call irreversible degenerative processes (such as arteriosclerosis, hypertension, and arthritis) "diseases" only because they accept as "normal" a way of life that breeds these afflictions? In short, how can they accept to deal at a merely individual level with the damages that our civilization and society cause to entire populations?

But, as soon as you have asked it, this question turns back on you as well. Why do you, wage-earners, citizens, voters, tax payers, constantly demand that the state or your immediate employers cover you against the consequences and the costs of illness, but not that they protect you against the illnesses themselves by eliminating the causes? Why do you incessantly demand more hospitals, doctors, nurses, new drugs, instead of asking for living conditions that would allow you to dispense with their "benefits" and services? Why, instead of changing your unhealthy habits and way of life, do you ask "your" doctor to relieve its effects?

Would you keep going to a doctor who told you squarely: "medical science can do nothing for you and furthermore if you could stop smoking, stop overeating, stop worrying, and stop spending your days sitting indoors, you wouldn't need medicine?" Would you even keep going to a doctor who claims to treat your flu as your grandmother did, saying to you: "Drink four quarts of hot lemonade a day, stay warm, rest, and you'll be cured in three days without medication"?

Come now, you know perfectly well the responsibility for over-consumption of medical attention and medication doesn't rest only on those who sell them by lying about their effectiveness, but also on those who buy them *and are just asking to be fooled*. This is what the book by Jean-Pierre Dupuy and Serge Karsenty—a book much richer than its title, *L'Invasion Pharmaceutique*, suggests—shows with great subtlety. The authors marvelously analyze the complicity between doctor and patient, neither of whom is entirely the dupe of the respective role.

Not that the patient is a faker and the doctor an impostor. It's much more complicated than that, for health and illness are also always a question of perception, and this perception varies according to the social and cultural context even more than it varies by the temperament of the individuals involved. The same symptoms will not be felt in the same way on Monday as on Saturday, before work as before a lovers' rendezvous. "Cultured" people, used to self-observation, feel sick more promptly than "unpolished" people, who are used to taking no notice. Wage earners who are frustrated by the stupidity of their fragmented work become ill more readily than farmers or craftpersons whose whole enterprise is in jeopardy if they do not finish their job.

Illness, as Dupuy and Karsenty remind us, is also a "strike" or a passive protest, and it is nothing more in most cases. General practitioners say that 75% of all patients have no organic lesion and come to the doctor looking *for comfort at least as much as treatment*. These sick people have no clinically definable disease, even though their troubles are real and can lead to organic lesions. Doctors call them "functionally ill" or "psychosomatic," and more often than not are willing to treat their symptoms with expensive and poisonous medications. That is where the fraud comes in.

In effect, these truly ill people who have no definable disease are most often *people who can't cope any more and come to ask for help and exemption of duty.* In another age they would doubtless have gone to confession, made a pilgrimage, or immersed themselves in prayer. Today we ask the laboratories rather than the saints to perform miracles. Charity is gone from heaven as from earth. To be socially acceptable, the cry for help must take the form of an organic disorder—exogenous and independent of the patient's will. You would have no chance at all of getting your boss or supervisor to listen to you if you said, "I can't go on; I'm losing sleep, my appetite, my interest in sex; I don't have any energy for anything anymore. Give me a week off." To be acceptable, your "I can't go on" must take the form of a somatic difficulty, of some impeachment beyond your control—in short, an illness justifying a medical exemption.

The person on the verge of breaking down therefore "somatizes" and "medicalizes" the dis-ease—not deliberately, but by the way that he or she feels it and interprets it—so as to bring in the only authority qualified to grant an exemption of duty—the doctor. And the doctor, in most cases, will play the game and treat as a chemically treatable illness what is basically merely the incapacity of the patient to bear the situation he or she has to face.

But this deception is fraught with serious risks, risks that avant-garde doctors saw well before Illich began pointing them out. This deception lies not only in the application of a technological treatment to a cry for help that appears in the form of an illness. More fundamentally it lies in the treatment of the patient's "I can't bear it" as a *temporary anomaly* that is medicine's mission to remove as quickly as possible. Here, then, we see how doctors and medicine are turned into agents for *social normalization*. Their mission is to eliminate the symptoms that make the patient maladjusted to his or her role and unfit for work.

Doctors will of course reply that they are beyond reproach, since "it is the patient who comes to us asking to be cured as quickly as possible." But this is a foul excuse. The patient's role, by definition, requires that he or she ask to be cured. The real question is "can medicine help the patient?" Rather than a temporary and in principle curable anomaly, might not the illness be the inevitable response of a healthy individual to a situation that is not? Aren't the digestive troubles, headaches, rheumatism, insomnia, and depressions that switchboard operators, key punch operators, assembly line workers, and electronics solderers suffer from, more than anything the "healthy" protests of an organism that cannot adjust to the violence done to it daily, at an eight hour stretch?

To act, then, as if it were the symptoms that are the evil rather than the work that causes them is to ask medicine to complete the job begun by the school, the army, the prison—that of producing individuals who are well-adjusted (by chemicals, if necessary) to the social role that society has cut out for them.

This is not an exaggeration. This is precisely the view of medicine held by conscientious, aging, or overworked employees who ask the doctor for the tranquilizers, stimulants, antidepressants, sleeping pills, etc. that will help them bear a role that has become unbearable—until they collapse. Again, this is how many company doctors see medicine, caring only to get the workers back as quickly as possible to the very job which makes them sick. Above all, this is how the function of medicine is seen by army, prison, asylum, and police doctors, who do not hesitate to "treat" the individuals who won't adjust to their confinement. There are drugs to calm the "restless," others to turn the "violent" into terrorized sheep, others to make homosexuals impotent, still others to keep those under torture from fainting or dying.[38]

At the end of this road there is compulsory psychiatric

treatment—or "brainwashing"—for the deviant, the maladjusted, rebellious, malcontent, lazy, etc. In the "best of worlds" not to be happy is to be sick. And sick people are to be treated. It is not only Soviet police and psychiatrists who think this way. They have illustrious colleagues in western Europe and the U.S.—for example Professor B.F. Skinner, whose methods of "reeducation" were faithfully depicted in *A Clockwork Orange*. Or the Hispano-American Delgado, who dreams of a world council of psychiatrists who will program all "normal" behavior from a distance via implanted computers—beginning with government leaders.[39] Or Professors Gross and Svab of the University of Hamburg, whose methods of personality destruction proved terrifyingly effective on German political prisoners.[40]

When therapeutists are willing to treat symptoms without wondering whether these "morbid" symptoms come from an organic disorder or from unacceptable conditions society imposes, they easily become auxilliaries of the police and the government.

So it is high time to rethink medicine or, more precisely, the determinants of health and illness. Illich's goal is to stimulate this rethinking. He is intensely concerned that medicine and society not respond to medicine's failures by treating evil with evil—by further enlarging the medical establishment, its jurisdiction and powers, and its tendency toward social control and the "medicalization" of life. According to Illich, the only healthy response to this crisis is a deprofessionalization of medicine, that is, the abolition of the doctors' monopoly in matters of health and illness, and the recovery of ordinary people's autonomous ability to take care of themselves. This point of view is not technically unrealistic (even though it requires radical politico-cultural changes).

The technical effectiveness of medicine is very limited. Hospitals could release 85% of their patients without harming them from a strictly medical point of view.[41] In three out of four cases the advice of a general practitioner, and the inevitable prescription that extends it, has a psychological (or psychosomatic) effect, not a technical one. That is, it has the same kind of effect that incantation, prayer, or exorcism had in the old days. In 75% of all cases the usefulness of the prescribed medication is not in the active principles but in the *faith* that the patients have in the technology. In other ages people *believed* in miracles; today they *believe* in science, and so the medical ritual takes on the appropriate guise. What is the difference between the doctor who reproached Powles for not having given antibiotics to a patient who "deserved them," even though they would be completely useless in her case, and the sorcerer or faith healer?

What remains are the 25% whose illness can be precisely diagnosed. Do they all need the technical care of a professional? Not at all. In 90% of all cases the illness gets better by itself. In these 90% the main purpose of the prescription is to order the rest, diet, and behavior—disguised as drops, pills, and suppositories—that will allow the patient to recover. In the end, there is only a very small proportion of sick people who need specialized care.

These are figures that put things into perspective. They indicate how much of the professional medical apparatus will remain technically necessary when medicine is stripped of its myths, its mysteries, its magic rituals. These figures show that deprofessionalization of health care is possible, and not only in China. "The overwhelming majority of diagnostic and therapeutic interventions which demonstrably do more good than harm have two characteristics," writes Illich, "the material resources for them are extremely cheap, and they can be packaged and designed for self-use or application by family members."[42]

This deprofessionalization of medicine, Illich notes, "is not meant to deny the training and skill of experts whom people may need on particular occasions." But it means that recourse to professionals should be occasional and kept to a minimum. For the society that offers its members optimal health is not the one that hands them over to a giant conglomerate of professional therapists. On the contrary, it is the one that "distributes among the total population the means and the responsibility for protecting health and coping with illness."

"Healthy people need no bureaucratic interference to mate, give birth, share the human condition, and die." Healthy people are not well medicalized people, but "people who live in healthy homes on a healthy diet, in an environment equally fit for birth, growth, work, healing, and dying, and sustained by a culture which enhances the conscious acceptance of limits to population, of aging, of incomplete recovery and ever imminent death."[43]

All cultures previous to ours, Illich reminds us, endeavored to agree to these inevitable and necessary limits. Health care was not the exclusive speciality of professional technicians. On the contrary, the art of staying healthy was the same as the art of living, with the rules of right conduct and "hygiene" (hygieia) in the original sense of the word. These rules were particularly concerned with "sleeping, eating, mating, working, playing, dreaming, and suffering," and they made people "able to bear pain, to understand illness, and to give meaning to the ever looming presence of death."

With industrialization the art of living ("hygiene") ceased to be built into all social activities. It is easy to understand why. As wage labor became widespread, workers stopped being in control of the length, intensity, pace, and conditions of their work. Unlike master craftspersons and landowning peasants, they are no longer able to regulate their work, their rest, and their sleep according to their own needs. Deprived of power over the rhythm of their lives, they are also deprived of *the culture and "hygiene" of work.*

Work thus becomes an external obligation that workers are forced to do under constraint. They tend to abandon the factory on the first pretext as soon as they can. The 18th and 19th century bosses made much of their "idleness." Obviously then, these "idlers" can't be trusted to decide for themselves when they are sick and when fit for work. This decision (a certificate of illness, or a certificate of fitness or of cure) must be made by specialists, who are applying "scientific" criteria. The growth of the clinic at the beginning of the last century made these criteria possible and illness became an entity separate from sick men and women, their work, and their lives. Rising capitalism took hold of these discoveries. From now on only the doctor will have the right to judge who is sick and who is not. Even the most ordinary troubles will become the object of medical care and certificates. In this way, capitalism came to dispossess people of their sickness and their health as it had dispossessed workers of control over their labor.

Since then, instead of being defined as a general condition of well-being, health has become simply a condition of non-illness, that is, physical fitness for work. The disease, on its side, stops being a condition of the sick person him or herself and becomes an "abnormal" hindrance which he or she has to get rid of as quickly as possible. From now on it is the disease that is studied, cared for, or cured, not the sick people. The introduction more than 100 years ago of health insurance gives further impetus to the professionalization, the industrialization, and the standardization of care.

Although Illich never puts it in these terms, his argument leads to the conclusion that the recovery of health would require the abolition of forced wage labor. It would require that workers regain control over the conditions, the tools, and the goals of their common work. It would require a new culture whose productive activities cease to be external obligations, so as to recover their autonomy, their diversity, and their natural rhythm...and so become joy, communication, and "hygiene," that is, the art of living.

It is necessary, Illich thinks, to demedicalize health just as it is necessary to deschool access to knowledge. For just as we won't recover a culture unless it is torn away from the school (so as to become the all-pervasive possibility of learning, teaching, and creating, whatever our age and our work), we won't recover health unless it stops being the business of specialists and becomes an endeavor and a value which are relevant everywhere, and which order individual and collective life at all times.

I know that it is not easy to follow Illich when he asks everyone to reject medicine individually. Taken literally this would suggest that wage earners give up sick leave and maternity leave. In fact, a healthy, demedicalized relationship to sickness and health will only be possible when we have abolished, along with wage labor, the "unhealthy" relationships (sustained by medical institutions and industries) that form the fabric of current society.

But Illich has no difficulty responding to this objection. The abolition of "unhealthy" social relationships can only be the work of women and men who, even within the framework of this society, will already have translated into standing rules of conduct their desire for the sovereignty of individuals and groups, the healthiness of our environment and way of life, and the establishment of relationships founded on fraternity and mutual help.

21 and 28 October 1974

3. Science and Class: The Case of Medicine

Private hospitals for the rich, and city hospitals for the poor; unequal access, according to social status and income, to equipment, medication, and the most expensive treatment; shortages of kidney machines, hospital rooms, scanners, and intensive care services; assembly line consultation for workers (three minutes a person, including the medical history, in Germany), as opposed to "personalized consultations" of half an hour or more in residential neighborhoods: this is still what the institutionalized left means by "class medicine." I do not intend to discuss it here— or at any rate very little. Not that discriminations of class, status, and income don't exist. But their existence is more the doing of individual doctors than of the medical institution and system.

True, there is still one medicine for the rich and another for the poor (just as there are diseases of the rich and of the poor). But this is true only insofar as there are doctors for the rich, and not because of a duality in the system. If it is more difficult for the poor to reap the advantages of advanced medicine and extravagantly expensive drugs, it isn't because the system rejects them. It is simply that since, for social and cultural reasons, they are more submissive to the authority of "their doctor," they are slower to dispute "him" and to find a specialist who will submit them to the most advanced technology.

For all that, can the rich get themselves more appropriate treatment, and are their doctors any better? Is there any reason to think so? Aren't there as many reasons to think the contrary? Wouldn't there be a better chance of finding charlatans among the doctors of the rich? Aren't they the ones who charge high rates for diagnosing diseases that defy classification, and who prescribe expensive drugs whose effect (if one may say so) is purely psychological? The privilege of having the death agony prolonged by two days or two weeks by means of technical feats is effectively reserved for those "important people" whose families the great master must convince, at their expressed or tacit request, that "we've done the impossible." But isn't this privilege a degrading torture both for the dying and for those caring for them, and not an enviable favor? Who says that because the rich pay for more extensive and costly medicalization they are in better health than anyone else? Isn't this merely the illusion that what costs more is better? Do you know which is the only social class whose physical and mental health appears to be superior in any measurable way, according to a still unfinished study by Dr. Brunetti? Country people of both sexes.[1]

Yes, I know, manual laborers and semiskilled workers die ten years younger than captains of industry and commerce. But who says that medicine and "health expenditures" cause the longevity of the rich? Those who live the longest, according to classification by profession, are school teachers and priests. Their longevity is not due to medicine.

Medical Treatment Doesn't Ensure Health

Still, one thing is certain: we do have class medicine. Only, class characteristics are not what we think. Take the fact just mentioned: the unequal life expectancy of different "professions." Has medicine ever shown a systematic interest in eliminating any of the causes that lead to the early deaths of manual laborers, or in making general the conditions that cause the longevity of schoolteachers and priests? Would you say that's not the role of medicine? Then we must wonder what exactly medicine is. Most doctors would reply that "medicine is the set of sciences that relate to human biology." Consequently, its object is to identify the agents of health and disease, to promote the maximization of the former and the minimalization of the latter. This has been the conception of medicine since Hippocrates.

If we keep this definition, one important practical conclusion must flow from it: medical science—meaning the knowledge of the agents of health and disease—cannot be fully effective unless the professionals who incorporate this knowledge do not remain its exclusive custodians. It's quite obvious; if doctors, because of continual study, are the ones who are the best acquainted with the agents of disease and the factors that make for health, then these agents cannot be eliminated or avoided and the factors cannot be combined in the best way—unless everyone is acquainted with the basic rules for a healthy life, and unless these rules (though variable within certain limits) enter into popular culture and lifestyles. This integration of medical knowledge into the culture, that is, into the art of living—meaning the art of working, of pacing one's days, of communicating with each other, loving, bringing up children, taking care of the old, of cleaning and dressing a wound, treating indigestion, feeding oneself, breathing, keeping oneself clean, eliminating wastes, watching out for the quality of the water and air, etc.—is what was originally called "hygiene." Knowledge of the conditions that make for health can only operate fully if it can be and is translated into "hygienic" activity that people practice *on their own* in order to keep or recover health.

Seen from this angle, the most striking victories Western medicine has won in the past 50 years are, more than anything else, advancements in hygiene. We have become accustomed to eating more varied and abundant foods, especially milk products and fruits and vegetables in all seasons; we have rebuilt most of the slums and shantytowns, exterminated vermin and rats, provided sewers and properly treated drinking water to everyone, taken to airing our rooms and workplaces, to using toilet paper, to washing our hands often, etc. All of this progress in hygiene has of course gone along with the development of the therapeutic apparatus. Nevertheless, it is less thanks to hardware and medical treatment than to the progress of hygiene that the general health of the population has improved spectacularly.

Although the very real progress of the therapeutic apparatus has made it possible to take better care of people who get infectious diseases, it is not by therapeutic means that the number and seriousness of epidemics has been diminished and that some diseases have completely disappeared, while others have become much less frequent and serious.[2] It is not because medicine knows how to treat a disease more efficiently that fewer and fewer people get it. It is rather the reverse that is true: an effective treatment can only be invariably successful when the disease has lost its endemic character. And it loses this thanks not to curative treatments, but to the elimination of the social, economic, ecological, and cultural causes of morbidity.

This is obvious for all the vitamin deficiency diseases, all the parasitic diseases, and for the great majority of the infectious diseases. All of these, with rare exceptions (of which poliomyelitis is one) first attack people who are already weakened from undernourishment, overwork, and unhealthy living conditions.

The Agents of Health and Disease are Primarily Social

The superiority of hygiene over treatment was the subject of a statistical investigation by the American epidemiologist Charles Stewart. On the basis of statistical comparison, 85.5% of the differences in life expectancy around the world can be explained by two factors: the piping in of drinking water and literacy. It goes without saying that these two factors never exist in isolation, but are also indicators of the general progress of hygiene and "welfare."

In France a series of statistical comparisons done in 1974 indicate the following agents of improved health. A rise of 10% in the population density of doctors lowers the morbidity rate by 0.3%. A reduction of 10% in the consumption of fats reduces mortality by 2.5%.[3]

Still more striking are the results of an inquiry by John McKnight into the main causes of hospitalization, done in a poor neighborhood in Chicago with 60,000 residents. These causes are, in order of importance: traffic accidents, muggings and violence, veneral disease, non-traffic accidents, bronchitis, dog bites, etc. On the whole, 75% of all hospitalization has a social cause.

Thus the epidemiology and biology of populations, which are part and parcel of medical science, very clearly attest to the modest role of curative medicine and the important role of environment, lifestyle, and hygiene (in the extended sense) in the struggle against morbidity and for improved health. However, Western medicine *as an institution* remains stubbornly impervious to the teaching of medicine *as a science*. In the West curative medicine continues to develop incomparably faster than hygiene and social prevention, which are given short shrift.[4] It is in this fact that the class character of medicine is revealed.

There are few models that illustrate so eloquently the fact that scientific knowlege will be neglected and almost censured if it does not conform with the interests and ideology of the dominant class. Worse still: those who censure and neglect the teachings of the epidemiology and biology of populations are also the ones who bear the responsibility for giving medical science its institutional form and who enjoy the monopoly of its institutionalized practice. No science exists independently of the institutions that ensure its transmission and practical insertion into the established order. Also, when I speak of "medicine" without further qualification, I don't mean "the biological science of humankind," but rather that which medicine really is: an institutional practice that selects both the possible applications of scientific knowledge and the knowledge itself so as to render them compatible with the prevailing social relations and the dominant ideology of capitalist society.

Thus our medicine is a bourgeois medicine in three principal ways:

1. It considers health and disease to be individual conditions, accounting for them by natural or accidental "causes" whose social dimension is concealed.

2. It favors individual consumption of supposedly health enhancing commodities and services, at the expense of all other health enhancing factors which it prefers not to recognize.

3. It favors the rare 5% of diseases that require very specialized care and expensive and complex equipment over the 95% that are the most common; it consequently ranks medical knowledge in a way that gives specialists of rare diseases the highest status and income. I shall return to this point.

Science and Institutions

The main factors of morbidity in our society are beginning to be well known and evaluated. Only "medicine" obstinately ignores them. In particular, we know that intestinal cancer is linked to a diet that is too poor in bulk, that stomach cancer is linked to particulate air pollution, that breast cancer is linked to a fatty diet.[5] We know that cardiovascular diseases are due to overeating, sedentariness, and stress. On the other hand, thanks to a recent American study, we know that members of certain religious sects, who practice a frugal way of life, take a lot of exercise, and live in well-integrated and stable communities and families, get half as many cancers and cardiovascular diseases as the rest of the population to which they belong. We know, although we are still unable to measure it, the pathogenic nature of air and water pollution by heavy metals, and of food pollution by pesticides, fertilizers, antibiotics, and hormones. We know that the postures that factories and offices force on workers are at the root of most "rheumatic" and circulatory diseases; that night shifts, noise, and the stress imposed by piece rates and assembly line work are the main causes of nervous and digestive troubles. We know that for nine million manual workers, *we record annually in France* 1,100,000 work-related accidents which incapacitate the involved workers for an average of 26 days. We know that, beyond accidents, the environment and nature of work in the United States (the only country where an official statistical evaluation has been done on this subject) causes 100,000 deaths and 390,000 disabling accidents a year.

Epidemiological research and inquiry teaches us then that the main causes of our diseases are social and that to eliminate them individuals must organize, inform themselves, and get their living and working places under their control, as well as their housing conditions and transportation—the things they consume and produce. Medicine ignores the need for this social approach. It carries on as if the only agents of disease that it cares to recognize are those it can fight without calling into question existing conditions. It makes a big deal of the chemical war against infections, of surgical skill, kidney machines or intensive care units that are sometimes able to save a few people who are seriously stricken. In its battle against the causes of disease, *it only wants to know those that a doctor can fight at the level of the individual organism, without getting into the social, economic, and cultural determinants.*

Doctors usually answer that by the time they see the patient the evil has already been done. They can only treat the man or woman at the individual level, with the techniques of the profession. They can't change the patient's job, nor the working and living conditions, nor the way of life. Undoubtedly. But this is true only at the level of the relation between the *individual* doctor and the *individual* patient. What, if not bourgeois ideology and social relations, is hindering medicine (and doctors as the technicians in the fight against disease) from going beyond the sphere of individual relations?

Why, with the exception of small health information groups, are doctors not organized at the local level, so as to act and call the population to fight all that runs counter to the requirements of hygiene and public health, against industrial pollution, against the pathogenic working conditions whose ravages they see in their patients? Why is it that doctors' associations do not set themselves up as public advocates for food hygiene and as public prosecutors of the chemicalization of food, agriculture, and cattle raising? Why does *medicine* accept with equanimity the habits of smoking and overeating whose evil effects doctors are satisfied to correct with another evil: the overconsumption of drugs?

Why? Fundamentally because *the practice of medicine is a business*. The relations between medical professionals and the public are market relations. The professional sells what the patients ask for or are willing to buy *individually*. No group of users of medical techniques is appealing to a doctors group with a view to action against the conditions that affect them as a group. Bourgeois social relations, and especially market relations, thus determine the way doctors conceive of their role and how medicine approaches the problem of disease, and of its causes and cures. And medicine, far from rising up against the ways that social relations deform and curtail medical techniques and knowledge, is in fact one of the strongest bastions of these social relations. Neither the Conseil de l'Ordre (the French version of the AMA) nor the hospital-university establishment will allow a group (such as a union chapter, a neighborhood group, or a consumers' group) to invest a doctors' group with a role of public defenders of public hygiene or expert witnesses for the indictment of night shifts, overtime, transportation conditions, food and drug industries, etc. *Medicine* holds its "impartiality" as the basis of its "scientific" credibility, and like all institutions that take part in the established order, it interprets "impartiality" as acceptance of the dominant norms and the power of the dominant class.

"Social Normalization" Through Medicine

In this connection it is not an exaggeration to view medicine as a particularly efficient and impressive machinery for the social normalization of individuals, and hence the repression of deviance and rebellion. Whenever medicine claims to treat or even cure the diseases that are the hardest to pin down, as if they were internal disorders which could be set right by a chemical assault on the organism, medicine is in fact acting in defense of the status quo. It implicitly imputes the illness to the sick organism and not to its living and working environment, and in so doing throws out as a possible cause the nature of the life and work against which the organism is rebelling or defending itself. Most illnesses, in fact, also mean "I can't carry on," an inability to adjust to or face any longer circumstances that involve physical, nervous, and psychic suffering—suffering that is unbearable in the long run for this person, or even for any healthy person.

When an electronics solderer, for example, suffers from headaches, dizziness, loss of appetite, and depression, is she suffering from a disease that she needs to be cured of? Certainly not. This worker (and most of her co-workers) *suffers from her work*, and it is this work that must be fought or abolished, not the morbid symptoms it brings on. These are merely the healthy responses by which the organism defends itself against the unbearable attacks of the work process. When medicine undertakes to suppress, or to relieve with drugs, the symptoms of suffering brought on by a pathogenic situation, it is fulfilling a repressive function. It is smothering an organic protest in order to get the "sick person," the maladjusted one, the "abnormal," readjusted as quickly as possible to the established order of things. And pushing this unadmitted logic of the medicalizing process to its conclusion, all deviation, distress, protest, or revolt can be taken for a pathological sign against which medicine will be called on to intervene. Psychiatric commitment of resisters does not occur only in the USSR.

The deepest logic of this medicalizing process can only be understood if it is put in the context of the spread of market relations and in particular wage labor. Hygiene, the art of healthy life, cannot be integrated into daily behavior and activity except insofar as people are masters of their own rhythms and their own environment of work and living. When they are urbanized and subjected to "compulsory wage labor," they lose control over their housing, working and living conditions. The possibility of "hygiene" is denied them, and their health is attacked at its cultural and existential roots. This is how it came to be that clinical medicine was born at the same time as industry. Clinicians set out to identify and classify diseases as entities independent of the sick person and requiring specialized medical care, thereby making it possible to diagnose illness or health without relying on the truthfulness of the patient's complaints. Clinical medicine was an indispensable complement to wage labor. When an individual sells his or her labor power wholesale to a boss and accepts a fixed wage in exchange for all the work he or she is strong enough to perform, this individual can no longer be trusted to decide when he or she is ill and when fit for work. Determination of the limits of his or her strength cannot be left to the worker, who is always suspected of malingering. Medical authority has to be called in to decide according to supposedly scientific criteria.

The same process that dispossesses people of their means of work, of their product, and their work skills, also dispossesses them of health and of illness. In the same way that they have to give up the free use of their labor power to a boss, they have to give up sovereignty over their bodies to medical authority.

At this point the doctor/expert's "science" becomes the *ideological* cover that legitimizes the bosses' authority: the power of the bourgeois class. As Ivan Illich writes, "medical diagnosis is an easy way of blaming the victim. The doctor, himself a member of the ruling class, judges that a particular person can no longer fit into the environment that was designed by other professionals, instead of accusing these latter of creating places to which the organism cannot adjust."[6] Having been obliged to abdicate to engineers the control over the use of their labor power, wage earners must then invest the doctors with full powers over their own bodies. Doctors alone "know" who is sick and who is not, who needs care and who does not. The doctors' submission to "science" conditions people to submit themselves to "those who know" and to delegate all their powers to experts. As John McKnight notes: "The more they believe that someone else should be in charge of their medical requirements, the more people behave like customers and the less they behave like free citizens. The customer relationship consists in believing that everything will work out better if you refer it to someone else who knows better than you do yourself."[7]

Submissiveness to medical authority and submissiveness to technocratic authority go hand in hand. There is no other country that requires as many vaccinations as France, and no country where centralized administrative power over its "citizens" is as pronounced.[8]

Professional Ideology vs. Social Usefulness

The professional organization of Western medicine has an elitist structure. It prizes rare diseases and costly technology, to which only a minority can have access, and it neglects the simple and inexpensive techniques of hygiene—techniques that would definitely be effective against the everyday afflictions that constitute 95% of all illness. It appropriates disproportionate resources for hard technologies and heroic interventions, and remains serenely powerless against the most common afflictions (colds, flu, "rheumatism," asthma, etc.), as if, because of their commonness, they were too "trivial" to warrant the profession's interest.

This indifference of professional medicine regarding the struggle against the most common ills is easy to understand. Endemic illnesses will not be effectively fought unless the preventive and curative measures are trivialized (like contraception, pregnancy tests, and "hygiene") to the point where they are within reach of every person and every group. But such measures would be advancing hygiene, which is popular culture, to the detriment of medicine, which is high culture. They would attack professional medicine's monopoly over health and illness. Discoveries connected to hygiene do not ensure either power or glory or wealth to those who make them (and this is undoubtedly why they are more often made by biologists rather than doctors), while heroic medicine fits in well with therapeutic ideology, which promises a dependent population that it will be increasingly taken care of by "those who know."

But there is more. The profession's internal ideology and hierarchy attaches much higher value to highly technical performances done in exceptional cases than to in-depth work against the most widespread ills. It is as if a doctor's professional value were acknowledged by the medical profession in inverse proportion to social usefulness. The same, moreover, is true for all the other scientific professions. In the same way, agriculture puts the traditional farmer's know-how at the very bottom of a value system that features at the top specialized geneticists and chemists (who precisely because of their specialization poorly estimate the consequences of their inventions—which are devastating in the long run). In the same way, astute mechanics, without whom nothing would run, are placed at the bottom of a pyramid whose peak is occupied by research engineers. And the general practioner, the nurse, and the "barefoot doctors," who, mingling with the people, are (or could be) the best able to spread effective health care and hygiene, are scorned by the profession, which prefers to give its highest status to the hyperspecialized mandarin who can diagnose the exceptional case that only comes up once in a million times.

From the point of view of the profession, the most socially useful medical workers are rendered ordinary and interchangeable by their numbers and the unexceptional character of both their skills and the illnesses they treat. *They further health but not science.* On the other hand, the hyperspecialized mandarins, who are custodians of a necessarily rare skill since the illnesses they study are exceptional, *advance science and thereby perpetuate the monopoly and power of the profession.* In so doing they occupy the top of the professional pyramid, even if they contribute nothing to the improvement of people's health.

This contradiction between the hierarchy of professional values and the extent of social usefulness is at the root of the distortion and the unequal development of knowledge. Left to itself, any closed profession tends to give itself mandarin structures and to place its self-reproduction, the perpetuation of its privileges and power, above all other ends.

Overcoming this contradiction requires a permanent struggle not against advanced research itself, but against mandarin ideology: an ideology which claims that the holders of exceptional knowledge are responsible to their peers alone, and not to their neighbors, to the people.

Lumière et Vie, no. 127, April 1976

NOTES

INTRODUCTION, MEDICINE AND ILLNESS, HEALTH AND SOCIETY

1. Ivan Illich, *Medical Nemesis: The Expropriation of Health* (New York: Pantheon, 1976). Jean-Pierre Dupuy and Serge Karsenty, *L'invasion Pharmaceutique* (Paris: Le Seuil, 1974).

2. The idea of health put out by decadent capitalism's media sub-culture is one of unlimited capacity for consumerist enjoyment. All fatigue, failure, discomfort, surfeit, non-conformity, or sorrow is resolved by medications. Most amphetamines prescribed in the United States go to women who want their appetites reduced so as to get thinner.

3. See *Takonala santé*, 1 rue des Fossés-Saint-Jacques, Paris. Editorial in no. 8.

4. Mainly cardiovascular diseases and cancer.

5. Warren Winkelstein and Fern E. French, "The Role of Ecology in the Design of a Health Care System," in *California Medicine*, 12 November 1970, pp. 113-117. See also René Dubos, *Man Adapting* (New Haven: Yale University Press, 1965).

6. Reproduced by John Powles in *Science, Medicine and Man*, vol. I, (London: Pergamon Press, 1974), p. 7. First published in *Antologia Medicina*, t. 7/4, CIDOC, Cuernavaca (Mexico), 8 volumes.

7. Letter from John Cassel, professor of epidemiology at the University of North Carolina, to the American Sociological Association, 29 August 1973. Reprinted in *Antologia Medicina*, t. 8/1, CIDOC, 1974.

8. Charles T. Stewart, "Allocation of Resources to Health," in *The Journal of Human Resources*, VI, 1971.

9. In *Science, Medicine and Man, op. cit.*

10.

	Life Expectancy	Infant Mortality	Doctors per 10,000 Inhab.
Barbados	69	47.7	4.2
Jamaica	69	35.4	4.9
Costa Rica	67	65.0	5.4
Canada	71	23.1	12.2
United States	70	23.3	15.6
Argentina	68	59.3	16.4

11. From L. Lebart, CREDOC, June 1970. Cited in *L'invasion Pharmaceutique, op. cit.*

12. Obviously we can't deduce from this that these two factors can be introduced by themselves. Water purification and literacy (not to be confused with schooling) imply a social and cultural revolution.

13. J.T. Lamb and R.R. Huntley, "The Hazards of Hospitalization," *Southern Medical Journal*, May 1967. The study was made at North Carolina Memorial Hospital.

14. Quoted by Charles Levinson in *Les Trusts du Médicament* (Paris: Le Seuil, 1974).

15. See *The New York Times* of 22 May 1974.

16. An evaluation by Professor Montagne in *Le Monde*, 3 May 1974.

17. James C. Doyle, "Unnecessary Hysterectomies," *JAMA* 151 (5), 53-01-31.

18. *La Découverte de la Maladie*, Centre de sociologie européene. Cited by Dupuy and Karsenty, *L'invasion Pharmaceutique, op. cit.*

19. Bergmann and Stamm, "The Morbidity of Cardiac Non-Disease," *The New England Journal of Medicine*, May 1967.

20. Ralph Audy and Fred L. Dunn, "Health and Disease," in *Human Ecology*, F. Sargent, ed., North Holland Publishing Company, Netherlands.

21. Cited by Kruse, *et. al., Bulletin of the New York Academy of Medicine*, vol. 33, 1957.

22. Paul D. Clote and John McKnight, *Automatic Multiphasic Health Testing: An Evaluation*, Northwestern University, November 1973. First published in *Antologia*, CIDOC, t. 8.

23. See *The Sunday Times* of 22 September 1974. The same is true for the smallpox vaccination, which is not advised in Great Britain and in the United States, since George Dick's epidemiological studies established that it now involves more risks than the disease itself (which is on the way to extinction).

24. In *JAMA*, 16 November 1970.

25. From Gordon Siegel, director (in 1969) of the U.S. Public Health Service.

26. So early that only by *monthly self-examination* can it be detected in time. The same goes for cancer of the cervix, for which self-examination has still to be worked out. Only the women's movement can invent a technique and implement it.

27. Frank Turnbull, British Columbia Cancer Institute, in *The Canadian Nurse*, August 1971.

28. In the French countryside, the farewell ceremony, in the course of which the dying person announced his or her last will, only fell into disuse at the end of the last century.

29. See Victor Sidel, "The Barefoot Doctors of the People's Republic of China" in *New England Journal of Medicine*, 15 June 1972. Reproduced in *Antologia Medicina*, t.4, CIDOC, Cuernavaca, 1973. The reader who gets the impression in the preceding paragraph that I am stating the uselessness of all specialization and all medication is asked to reread it very attentively. It is about *common* illnesses and diseases.

30. The Chilean doctors who participated in this revolutionization of medicine were all assassinated the week following the military putsch of 11 September 1973. See *Medical Nemesis, op. cit.*

31. J.N. Morris, *Uses of Epidemiology* (Edinburgh: Livingstone, 1964). Cited by Powles in *Science, Medicine and Man, op. cit.* Reproduced in *Antologia Medicina*, t.7.

32. See W. Winkelstein and F.E. French in *California Medicine*, 12 November 1970.

33. Cited by Powles, *op. cit.* American epidemiological studies have recently established that while breast cancer is much rarer in Japanese women than in American women, it is just as common in Japanese born in the United States of immigrant parents.

34. See Lare and Saskin, in *Science*, 21 August 1970.

35. Geoffrey Rose, "Epidemiology of Ischaemic Heart Disease," *British Journal of Hospital Medicine*, 1972, pp. 285-288. Cited by Powles, *op. cit.*

36. The article by Jean-Pierre Dupuy is in the *Review of Political Economy*, January 1974.

37. John Cassel, in *Antologia Medicina*, t. 8/2. Emphasis added.

38. See the interview of Dr. Peter Breggin in *Liberation*, vol. 17, no. 7 (1972) and John Saxe Fernandez' contribution to *Sociology of Terror* (New York: Doubleday, 1974).

39. Jose Delgado, *Physical Control of the Brain* (New York: American Museum of Natural History, 1965).

40. See in particular the articles by Dr. Sjef Teuns in "Les Prisonniers politiques ouest-allemands accusent," *Les Temps Modernes*, no. 332, March 1974.

41. See the first part of this article.

42. *Medical Nemesis, op. cit.*, p. 120.

43. *Ibid.*, pp. 274-275.

SCIENCE AND CLASS

1. In the national statistics, the longevity of country people is nevertheless lowered by a factor that is independent of their work and the conditions of their life: endemic alcoholism is certain in rural areas. Cf., Alain Letourmy, *Santé, Environment, Consommations Médicales* (Paris: Cérèbe, 1974).

2. Nor thanks to vaccination. With the exception of poliomyelitis, infectious diseases, against which vaccination is compulsory, declined at the same rate before and after the introduction of the vaccination requirement; and, in countries with comparable civilization and standards of living, whether vaccination was generalized or not.

3. From Alain Letourmy, *op. cit.*

4. "Prevention" is limited to vaccinations which, according to epidemiological studies, have more disadvantages than advantages in "developed" countries, and to early detection whose risky nature and doubtful usefulness have been demonstrated many times. There is a bibliography on this subject in *Medical Nemesis, op. cit.*, chapters 2 and 5.

5. Cf., American Cancer Society, *Persons at High Risk of Cancer* (New York: Academic Press, 1976).

6. Ivan Illich, *Medical Nemesis, op. cit.*, chapter 7 (in the enlarged French edition).

7. John McKnight, "The Medicalization of Politics," in *The Christian Century*, 17 September 1975.

8. In the Anglo-Saxon countries the smallpox vaccination is nowadays not generally recommended, except for people who must go to places where the disease persists. The whooping cough vaccination is considered more dangerous than the disease itself in Great Britain and West Germany. In any case, the more or less effective protection of a vaccination is limited by time. Compulsory vaccinations can only have the effectiveness they are supposed to have if regular boosters are equally compulsory.

Epilogue

Continuing the American Revolution

This piece is not meant to be a comprehensive account of the situation in the United States. It attempts to convey to a French public some of the unique qualities that make the U.S. people positively different from any other and that would make it worth while for French intellectuals not to restrict their interest to U.S. women, blacks and homosexuals alone. I expect U.S. radicals to be as irritated by this piece as I often am by the hopes and trust they put in the French left.

Jim

In the dazzling sunshine we drive through a maze of seven freeways leading north out of San Diego. Jim is still over-come: this morning his father spoke to him on the telephone for the first time in four years. Jim's father is a retired colonel. He disowned Jim when he was arrested for the fourth time in 1972. At that time the students were preparing to use any means to prevent the Republican Party (headed by Nixon) from holding its convention in San Diego.

Nowadays Jim teaches off and on at the university. "Is your work going well?" his mother had asked him on the phone. "You're not in any trouble?" Jim had reassured her: she had nothing to worry about. That's when the colonel came to the phone: "You're lying, son," he said. "I saw it in the papers. Last week you were in a demonstration against the dean because he wouldn't give up projects paid for by the CIA. And the dean is threatening to throw you out. Well, kid, you just stand up against this son of a bitch. Give 'em hell! Give 'em hell!" And he hung up.

What had happened in the colonel's mind? Jim said simply, "Watergate." First of all Watergate, and then the revelations about the CIA, the FBI, Lockheed, etc. The legitimacy of the institutions has collapsed. Those who had believed in the authority of the state and in the patriotic duty to serve it have lost their illusions. The colonel shakes hands with his prodigal son. He is not the only one.

Susie

Susie has found a job after two years of unemployment. We are strolling along the walk built near the cliffs. Half the men are bare-chested; there are swimmers in the Pacific surf. Every 20 yards young men with long hair stop us: "Jesus loves you. Do you want to meet him?"

"Meet *her*," laughs Susie. "Jesus was a woman, didn't you know?" The women's movement has decided to take up arms against the masculinity of God: "Why He, God, and not She, God? Is theology sexist, or what? And religion, is that for men? They can keep it!" Susie is enjoying the surrealistic effect of this notion.

"Do you see that grey band there on the horizon? That's the Los Angeles smog. And the dark cloud in the sky over there? That's the San Diego smog. San Diego is the most spread-out city in the United States. Here the air is still clear and you can run on the beaches. But how long will it last? Oh, do you see that spray over there? That's a whale spouting. Oh, there's another. When I see a whale in the morning, I'm happy all day long."

George

Ocean Beach is just like a village. It has wooden houses with porches and little gardens, narrow streets without traffic, flowers and vegetables that the residents grow even on the earth strips of the sidewalks. Fifteen thousand manual laborers, officeworkers, and unemployed (20% unemployment) live in this San Diego neighborhood, modestly, but in luxury. "Their luxury is the beach," says George. "You spend your days there—you surf, fish, make love, or do nothing."

When the first developer came to put up the first four-story apartment block, they organized, occupied the site, fought with the police, besieged the city hall, and finally won. The town reversed the engines. When the building was completed, it couldn't be rented because the residents arranged to discourage prospective tenants.

There are thousands of stories like this all across the United States. People here always start with the idea that the country belongs to them—to them, and not to the government, the cops, the banks, the army, industry, or any of the authorities bribed by big business. The former militants of the heroic era of the Vietnam war have at last found in the neighborhoods a base that is both social and territorial and that is quick to resort to direct action. Almost everywhere you can see stanchions for elevated highways, stopped short, outlined against the sky—testimony to residents' organizations that kept the bulldozers out of their neighborhoods. In Palo Alto, near San Francisco, there are acres of subdivided land, laid out in asphalt streets and crossed by gas and water lines, where there will never be any houses. The residents (organized by a former SDS member) forced the city council to withdraw the construction permits.

At Ocean Beach the struggle has left some new things: a cooperative bookstore that carries all the leftist publications, a food coop ($500,000 worth of business), built and managed entirely by volunteer work and selling mainly organic products brought in from farming communes in the area, and a cooperative restaurant serving a complete meal for $1.40. The owner gave the restaurant to his workers and left to grow organic vegetables in the hills.

"What do you think about this?" asks George. "They say we are practicing socialism in one neighborhood. They say we have to build a party that can coordinate and unify the thousands of local movements." But some people have been trying to do that for five years and it hasn't gotten anywhere. A political party addresses people's opinions and traditions, a local movement their tangible experience. And to start a party you need money, organizational structures, delegates, full-time workers.

"As soon as you give people the impression that you want to tell them what to do, that's it. They go home. The only thing they care about is to govern themselves, right here and now, and not to have a better government in Washington." I object, predictably, that you can't dispense with taking over the government. "Obviously," answers George, "you are logically correct. But in fact how do you expect to go about it? By electing a candidate who will make radical reforms? Everyone knows perfectly well that if such a candidate were elected he'd be assassinated. Look, we seem to be a very free society; we have a lot of space here. And it's true that the government is afraid of the people. But the people are in their neighborhoods; they aren't in Washington. And as soon as you try to establish the power of the people in Washington, you discover that the government is a block of steel and concrete, or rather of bankers, cops, and military men. Legally there are unlimited possibilities for reform here. In actual fact, the space for reform is limited to the range of a rifle with a telescopic lens."

Heinz

In a few years, when he retires, Heinz will build his house here in northern California, not far from San Francisco. He will build it entirely with his own hands; it will have solar heating and hot water, a biogas generator, a greenhouse, and other devices of soft technology. Heinz talks with his whole body. In the true sense of it, he is an idea incarnate. This typical American is in fact an aristocrat of Viennese birth (that is, German-Jewish-Slavic). Heinz's name is von Foerster. Along with Norbert Wiener, John von Neumann, and others, he is one of the pioneers of cybernetics—the theory of self-regulating systems.

Shortly after the war, Heinz worked out a mathematical theory of physiological memory which he published in a pamphlet. While visiting the United States in 1949, he got a telephone message from the University of Illinois, 500 miles away. Some people there had read his pamphlet and were inviting him to come and see them. When he got there Heinz discovered that his hosts had experimentally worked out parameters for which they had no theoretical explanation; in contrast, Heinz had established these same parameters theoretically, but had been unable to support them by experimental verification.

He was hired on the spot. He didn't speak English? No obstacle. So that he could learn it, the group made him editor of their periodical. Heinz is still at the University of Illinois, where he has created the Biological Computer Laboratory. From what he told me I understood that biology was bound to eat up the other sciences, including economics. For the very nature of its object-life compels biology to break with the analytic approach that leads us to "know more and more about less and less" in favor of a "holistic" approach. A living organism is not a machine. In its essence it is autonomous.

Of course, this is what bothers many people. The dominant tendency in science as in society is to suppress autonomy in favor of external determinants. It means to "trivialize" individuals (Heinz calls machines trivial when they give identical, strictly predictable responses to a given action) by destroying in them this troublesome dimension called autonomy—this source of unpredictableness and novelty. What is required of the "good student?" To know the right answer? Or only to ask the questions to which the teachers have the right answer? How about teaching them to ask questions to which the answers have yet to be sought? For example, what do we need in order to trivialize society, but not individuals? Why, right now, don't we have the opposite situation—individuals are trivialized and their behavior is made statistically predictable? But the results of this are invariably contrary to individual goals. This society is a one-way system in which the media are continually speaking to people who can neither talk back nor communicate with each other. There is no feedback. This is why the system becomes complicated, oversized, bureaucratized, out of people's control.

I suggest to Heinz that he is working out in his own language the main issues of this "theory of practical ensembles," which is Sartre's *Critique of Dialectical Reason.* He wants to know more and listens with the same communicative intensity he puts into talking.

Dan

Dan checks his speedometer as he takes the left fork of the highway. The needle wavers around 60. The speed limit is 55. Three citations in a year (or five in two years), and you have to go to driving school or face a jail term.

Dan points out some blocks of wooden bungalows flanked by a sheet iron hanger. Those are the advanced electronics firms. No, the big electronics firms are not the ones at the forefront. Didn't you know that? These are the mavericks who miniaturized the calculators. Yes, yes, the "chips," the micromemories that store 2000 bits of information on the head of a pin. They are born in little bungalows like that one. No indeed—the inventions that count rarely come out of the big laboratories. Why would a little genius who had an idea go sell it to IBM? They'd buy the idea and put it away in a drawer. While if he puts it into production, even on a small scale....

Dan knows what he's talking about. Do you know the biggest firm specializing in closed circuit TV? That's him. How many employees? Thirty. And among those thirty are two little geniuses, almost completely useless to the business, who come in when they feel like it, draw a middle range salary for undefined work, and invent astonishing things. "Why do I keep them? For conversation." No, that's not a frivolous reason. These guys have good noses. They figure out the things that'll sell in three years or in five years. And when he has a really tough problem, Dan can consult them. For Dan is self-taught.

The vitality of the United States, Dan says, is in its entrepreneurs. The day they're gone we'll be like the USSR. Go ask the heads of the big monopolies what technological changes they foresee in the next 20 years. They'll blow you away with high-sounding formulas. The future—Dan points to a wooden building—is in there, if there is any. That's where solar power is beginning to emerge from. "People are starting to understand this new fact: capitalism and free enterprise are not only two separate things, they are two contradictory things. The day when Americans become convinced that 'socialism' doesn't mean less freedom and more bureaucracy, but exactly the opposite, there will be a tidal wave of socialism in this country."

Ralph Nader has understood this perfectly. In Europe he is still taken for a consumer advocate, although he has initiated a movement with branches in most of the big cities and universities. Nader's idea has always been that people have to organize and take power over their own lives—that is, over everything they can control directly. Obviously to do this you must get rid of the powers that be, starting with the economic powers. So Nader began by making the crimes of big business, and its collusion with big government, tangible to people.

Now Nader has to make his movement more clearly political; he has to propose an alternative to the system. He is turning toward a kind of self-regulating anarcho-socialism. He talks about structural changes, defeating the power of big business, multiplying the number of small cooperative production units, putting factories under worker control, and having consumers' associations control distribution so that inferior products wouldn't find any takers. He talks about the withering away of the state, about a "new kind of socialism, founded on local power." He wants to go visit China and Yugoslavia.

Jerry Brown

Jerry Brown's models are Ho Chi Minh, Ghandi, and Mao. His bedside reading is *Small Is Beautiful*, subtitled "Economics As If People Mattered." Jerry Brown, a former pupil of the Jesuits, is the governor of California, and he spends a lot of time at the zen (Buddhist) center. The most revealing chapter in *Small Is Beautiful* is entitled "Buddhist Economics" and it says:

"The modern economist measures the 'standard of living' using the assumption that someone who consumes more lives better than someone who consumes less. A Buddhist economist would say this assumption is the height of paradox; consumption being only a means to well-being, the goal should be to get the maximum well-being from the minimum consumption. Thus, if a piece of clothing is supposed to keep you warm and make you look good, the goal should be achieved with the minimum work and material.... Modernization has impoverished people both materially and spiritually" by inviting unemployment and dependence on a wasteful market production system.

The author, who is very well known in the English-speaking world, is named E.F. Schumacher. He is a former director of the (nationalized) British Coal Mines, and economic consultant to several Third World governments.

In trying to apply Schumacher's principles, beginning with himself, Brown has become immensely popular. He refuses to live in the governor's mansion, he sleeps on a mattress on the floor in a rented apartment in town, and he makes his staff go on work retreats that can last from 7 AM to 2 AM. Somewhat like Fidel Castro, he shows up where he's least expected and gets into his constituents' problems, asking seemingly ingenuous questions which in fact spring from a kind of Socratic irony. For example, "why does a janitor, who does an unpleasant job, make less money than a judge, who is lucky enough to have an interesting occupation?"

Jerry Brown's idea, like that of Nader and the neo-anarchists in another guise, is that the institutions created to take charge of people's lives (their education, health, livelihood, living arrangements, jobs, leisure activities, etc.) generate dependency and frustration, powerlessness and aggressive discontent, passivity and resentment. Rather than expanding institutions to take further charge of people, he means to enlarge people's own spheres of sovereignty—that is, the possibility of solving their problems independently. Jerry Brown's slogan is: expect less from government and more from yourself.

John

Like many radicals, John agrees with Governor Brown's philosophy while holding that he is a clever imposter. To be governor and oppose institutional expansion is to conciliate the conservatives and the Wallace-ites without at all enlarging the "sphere of sovereignty" of those who are most deprived. "The truth is," says John, "Jerry Brown's politics are only words. He is using neo-anarchist issues to put together an election platform and get to the White House in 1980."

John is professor of urban affairs 1500 miles from here. Last year part of his teaching involved an inquiry into the causes of hospitalization in a working class neighborhood of 60,000 residents. The results: 17% were traffic accidents, 10% were muggings (and violence), 7% were venereal diseases. After these came, in order of importance, household and work accidents, bronchitis from unhealthy housing conditions, dog bites, etc.

In all, 75% of all hospitalization was caused by social problems that do not require medical remedies but political ones. The inquiry was "participatory" and helped to mobilize the neighborhood. It effectively established or strengthened street and block organizations, and elections of delegates were held by block, by apartment house, by street, by neighborhood. Work was shared voluntarily, and a plan of action was drawn up. The goal: to force the city to have a health policy (and not facilities for caring for disease) whose goal would be the emptying of the hospitals. Slogan: "*We want fewer sick people, not more hospital beds.*"

The hospital workers union invited John to its congress. He spoke for an hour. The title: "The medicalization of politics." John told the participants that society disguises as medical problems issues that require political action. He said we use hospitals and their workers to conceal the real, sociopolitical causes of the evil. Moreover, from any point of view hospitals are abominable places where even the healthy risk the loss of their health. There is no way a hospital can restore or maintain health when it is permanently undermined by social conditions. Thus hospital workers are at one and the same time the instruments and the victims of a gigantic swindle.

There were 4000 participants at the congress, and John was very anxious about how they would react. Well, as soon as he finished, the room rose as one person, and John clocked a four minute standing ovation.

"And now," John said, "how can we go beyond the level of local politics? And what shall we do in order to take our fate in our own hands instead of simply making demands on the authorities?" Always the same questions. So far there have been three initiatives: certain streets are blocked off from traffic, then there is a student-unemployed workers coalition for rooftop construction of, first, a greenhouse and then a solar heater made of old tin cans. The solar unit can lower heating costs by 60% and the greenhouse can provide the fresh vegetables that the unemployed can't afford to buy.

"If you make unemployment bearable instead of fighting it, you encourage people to do odd jobs rather than to engage in political struggle." John has made this objection himself. He refutes it now. In the big city ghettos the unemployment rate is 20 to 30%. It is over 50% for young people. Unemployment passes from (unknown) father to son; people are doomed at birth to live on welfare. This has been going on for over 20 years. Is the best preparation for fighting the system to refuse to do anything by yourself? Does liberation mean the government takes charge of people's lives and solves their problems? Or is it the possibility for people to solve their own problems, at the level of the city, neighborhood, community, production unit, etc.?

"As long as people expect welfare to take care of them," says John, "the most you can hope for will be riots, not revolution." The Black Panthers understood this very well. You have to fight the expectation of external solutions as much as drugs, prostitution, theft, and the dog-eat-dog attitude. And the only way to do that is to give people confidence in their capacities to do something for themselves.

I argue that the Panthers had a national organization, a political identity that made them immune to localism. John agrees that a political identity is important—"to succeed a movement needs a name"—but thinks a national organization is counter-productive. It caused the Panthers' ruin. It was enough to infiltrate their general staff, to assassinate some of the leaders, and play the survivors off against each other for the (decapitated) movement to collapse.

The most widely read publication in the United States is a (mostly organic) gardening magazine, circulation: 15 million. The only Marxist publication in San Francisco prints 6000 copies, Palo Alto's pacifist sheet, 12,000. More than a third of all U.S. households bake their own bread, at least occasionally. Still, various supermarkets sell many kinds of organic breads.

There is an institute in Berkeley, the Farallones Institute, financed by private foundations, gifts, and federal grants, that teaches subsistance techniques, soft technology, self-reliance in both urban and rural milieux, how to build your own house, etc. Tuition fee: $750 for ten weeks, $1000 if you want to get a degree. The institute, affiliated with Antioch College West, is setting up an 80 acre pilot farm with its "integral house." It has an "integral urban house" in Berkeley. A house is considered "integral" when it supplies the food and heating of its inhabitants through a system of conservation, recycling, aquaculture, fish-breeding, etc. Only air, water, and solar energy are taken from outside.

Many universities have such a unit, where teaching, research, and practical work are carried on all together. The movement began in Washington, home of three of the best known groups—in particular Community Technology (CT), founded by Karl Hess in the Adams Morgan section (31,000 mostly poor residents). Hess brought a substantial number of the residents to subscribe to a communal workshop where anybody can come and build solar equipment, roof greenhouses, and even all you need to breed trout in your basement. The fish excrement is used to fertilize the vegetables in the greenhouse and to grow the algae that fatten the trout. Hess is a celebrity and his goals are overtly political.

208   EPILOGUE

Karl

Karl Hess left school at 15 for a meteoric career with one of the big broadcasting systems. When his boss found out his age, he was fired. He moved on to writing for newpapers. Half as a game he agreed to write speeches for the Republican presidential candidate. That was in 1948. The Republicans thought he was brilliant, and later Tom Dewey, Eisenhower, Nixon, and Goldwater hired him as a speech-writer. "If Goldwater had won," Hess says, "I'd have sent to prison the people who today are my friends."

Goldwater was defeated (by Johnson) and as the war in Vietnam kept escalating, Hess realized that there was no more connection between the defense of freedom and the bombing of Hanoi than there was between "free enterprise" and the interests of Standard Oil or the Morgan Guarantee Trust, but rather that the imperialist politics of big government were not unconnected with the interests of big business multinationals.

Hess then began to listen to the "new left," saw the light— "everything they said was true"—immediately joined SDS, and then went to work for the Institute for Policy Studies (IPS), an organization which covers all shades of leftist tendencies. Karl's research project was to show why the small units are more efficient than large ones, to determine the threshholds, and to assemble material for an encyclopedia of science and technology for human-scale communities.

"In all fields there are technologies and tools perfectly adapted to decentralized use (miniaturization, cybernation, alternative energy sources, etc.) and yet, because of capitalist organizing and government, they continue to be used on an increasingly gigantic scale..."

Why? To insure that people are dominated by the system. Which makes it necessary to show people how, even in the cities, they can break this domination by running things themselves. "This is much easier in the city than in the country. City people can start production and distribution cooperatives and community banks, and even a poor neighborhood can produce almost all its own food itself."

Karl undertook to prove this after his motorbike brought him to break completely with the system. It earned him the general disapproval of the residents in his middle class neighborhood. It taught him to hate hypocrisy and the values of the ruling class that he had naively served since 1948. It sent him to live among poor people. And there, in order not to participate in the exploitative money system any longer, he decided to reject all money and all money relations. He learned welding and asked people to give him what they could for his work, in kind.

Then with a group of scientists and craftspeople, Karl founded Community Technology, whose goal is "to show how an advanced technology can be made to serve the needs and imagination of an urban community directly... Each project allows engineers, technicians, and craftspeople to rethink their respective roles and to experiment scientifically with a new way of working. The scientists thus will be able to put into practice their opposition to the priorities and goals which capitalist organization assigns to science. The neighborhood groups, for their part, will be able to enact their rejection of the capitalist way of life and capitalist economics by working out the material basis for a different life style and mode of production—the basis for a non-hierarchical, decentralized society, founded on direct rather than representative democracy."

Crisis I

Capitalism has been discredited; big business is on the defensive, ideologically. The legitimacy of political institutions collapsed with Watergate. A Senate investigation has shown that Los Angeles had an excellent public transportation system before General Motors bought up the trolley lines and replaced them, first with buses, and then with expressways clogged with cars.

Doctors in California are going on strike because their business is no longer profitable—the insurance rates they must pay to protect themselves against malpractice suits are so high. The police are corrupt; the army, corrupted by the arms industry, is corrupting the Congressional Committee on Defense by bribery. The school system is bankrupt; the big cities and urban planning are bankrupt. For the first time since the beginning of the century rural areas are being repopulated.

"Our institutions are like dinosaurs. Trapped in their own enormous mass of bureaucratic red tape, they are incapable of the simplest tasks. Still, they continue to grow. But when the climate changes—and it is changing quickly—dinosaurs are unable to adapt to it and perish because of their sheer size. And in their place appear small organizations of people who have a sense of personal responsibility, who are able to deal with the new situation. I am part of a vast movement. It gains ground every time a factory closes and fires its workers, every time a new industrial product is added to the list of carcinogens. The revolution is not made by bigwigs like me, but by tens of millions of little people trying to organize to take control of their own lives." (Karl Hess)

This dysfunctioning of the institutions is the first of three levels of crisis.

Crisis II

Nicholas Georgesco-Roegen is still known to only a limited public. When he receives the Nobel Prize—simply a matter of time, says *The Scientific American*—everyone will know what this Rumanian-born economist and professor at the University of Virginia has been insisting for the past ten years: the inability of economic theory to incorporate ecological reality.

"To talk about the cost of nonrenewable resources or irreparable damage is nonsense.... One of the most serious ecological problems is how to distribute a stock of finite resources to *all* the generations to come. Economics manages these resources for *only one* generation." All economists, including those of the Club of Rome, have so far forgotten that a stock of limited resources inexorably runs out "even if consumption of them stops growing, and even if consumption diminishes."

But here's more. In three recent issues, *The New Yorker* (always on the lookout for subjects that will be fashionable next season) is running an advance publication of the whole of Barry Commoner's new book, *The Poverty of Power* [New

York: Knopf, 1976]. This biology professor was the first to show that the environmental crisis is not due to economic growth, but to capitalist growth based on maximization of flows. In his new book Commoner goes even further. He demonstrates the profound structural link between ecological crisis and the crisis of capitalism.

Commoner's brainstorm was to put side by side, sector by sector, value added per BTU of energy consumed and per dollar of invested capital. The stunning conclusion: the industries that tie up the most capital are also the ones that consume the most energy. For the same value added, the petrochemical industry, for example, requires eleven times more capital and twelve times more energy than the leather industry.

Now, capitalist development rests precisely on the systematic substitution of plastic for leather, synthetic fibers for natural fibers, concrete for stone, highway and air transport for rail and boat transport, synthetic fertilizer for natural fertilizer—in short, of products that require a lot of capital and energy for those that use little. In the same way, within each sector, capitalism uses increasing amounts of fixed capital to insure the same level of production; as machines supplant human labor, consumption of fossil fuels replaces the energy of the body.

Because of this elevation of the "organic composition of capital," predicted and described by Marx, only an equally increasing mass of profits will allow the production apparatus to reproduce itself—that is, to replace and renew existing plants. Now, says Commoner, citing impressive statistics from financial agencies, industry is no longer able to realize the necessary profits. Its rate of self-financing fell from 70% in the 1950s to 26% for the five years 1970-74. According to the Chase Manhattan Bank, industry will be short by $1500 billion in covering its capital needs for the next ten years. "The ability of the system to reproduce itself is in crisis," Commoner says.

What makes this crisis different from previous ones, he goes on, is that the classic crisis of overaccumulation is aggravated this time by two new ecological factors. These account for the paradoxical situation in which endlessly increasing quantities of capital are needed simply *to maintain* the current level of production. The two factors are:

- the exhaustion of the most accessible mineral deposits and thus the much higher costs of exploration and extraction;
- the necessity to reduce industrial pollution.

Under these conditions there can be no question of coping with the lower profit rate by destroying capital. The capitalist system, Commoner concludes, is sapped at its roots by its structural inability to reproduce capital and make it profitable according to its own rules. It only maintains itself by violating the principles by which it functions, most notably by inflation and financial help from the government.

This is the second level of the crisis.

Crisis III

From New Mexico and southern Texas to Kansas and northern Iowa the winter grain sowings have failed. The drought in the Great Plains is already worse than the one that began in 1934. Colorado and Oklahoma have lost two-thirds of their 1976 harvest. The agricultural and pasture lands of California have been declared disaster areas.

In a recent report, the National Academy of Sciences supports the idea that from 1880 to 1940 the climate was getting warmer and has been getting colder ever since. The agricultural season has decreased by three weeks in Europe and North America. In 700,000 years there have been only five periods as warm as this one, and they each lasted from 8000 to 12,000 years. The present climate has lasted 10,000 years. The sudden arrival of a new ice age is not out of the question. The drought in Europe, in North America, in the Sahara, in South Asia—where the monsoon has become uncertain—could be an advance warning of it.

But even small temperature variations can have catastrophic effects, for modern agriculture has staked everything on fragile strains of grain, strains that are gluttons for water and energy and cannot withstand noticeable climatic changes. Varieties that are hardier but have a smaller yield have been almost wiped out.

Summers without a harvest were not exceptional before

1880. A third of the Finnish population died of hunger in 1693. In the same decade Scotland went for six summers without a harvest. The riots of 1709 in Anjou, Rouen, and Paris were linked to famine, which was killing people all over Europe.

Many Americans are saying that it is realistic to consider a return of the great famines, to be preceeded or followed by the collapse of current institutions. Ownership of a bit of land is considered the best protection against this eventuality.

This is the third level of the crisis.

Lee

It doesn't matter much to Lee how it all ends. What is clear is that it cannot go on. The industrial age (and not just growth) is coming to its end, and in its death agony the hundreds of communities Lee is in touch with will be the nucleus of a new civilization, one that is based on a well-balanced relationship between humankind and nature.

Lee is 28. He was jailed eight times during the Vietnam war. He lives 35 miles north of San Francisco in a town of 30,000 inhabitants in a little house with a tiny vegetable garden, two goats, a few chickens and a lot of books. He is the link between dozens of communes, groups, and cooperatives that he advised and helped to get started. Here, a cooperative car repair workshop is struggling against great odds ("People bring their psychological problems along with their mechanical troubles"). There, a genuine marketing cooperative: it has 500 members ("more than that gets into bureaucracy and anonymity"), charges 50¢ a week subscription fee ("too high a subscription encourages consumption"), and requires each member to donate two hours of work a month. At the check-out counter each person hands in the list of his or her purchases and their prices. The co-op employs three half-time people to run it. There is a waiting list of more than 300 applicants. Lee thinks it's time to start a second cooperative.

The development of the "movement" lies not in the growth but in the propagation of autonomous groups. "If you want to recruit, join the army." A group should never get so large that people are unable to talk to each other in a conversational tone, and to control, understand, and decide on all jobs in common. No power structures. If you attract too many people, divide the group. Help to create new groups that will be as much in control as yours is of their own work, initiatives, and production.

Between the agricultural-crafts commune and the weekend garden there are all sorts of intermediate stages. For example, half-time work, or job-sharing. That is, a full-time job is shared among two or three people who prefer to spend most of their time with their children, their animals, and their gardens, and who are satisfied with a half or one-third salary. Lee hopes to make job sharing legal. He doesn't think that this would reconcile people to a system in which unemployment is chronic. "On the contrary. We are showing that you can live better with less, that real satisfactions are found outside the system."

From our discussion I remember this: "In this country," says Lee, "nothing prevents you from being fully human. What I mean is, material poverty doesn't force you to struggle with your neighbors for the necessities of life. If people do not act like human beings, it is because they are prisoners of a system, of a mentality, and of an ideology that was forged in the struggle against scarcity. We will never abolish this system unless we act according to a different mentality and ideology, that is, unless we behave from now on according to our own idea of what a human being should be like. You say they'll lock us up or kill us? Then they will only show that they themselves are not human. Without us no one would even have noticed."

I argue with Lee that the non-humans are likely to be insensitive to his demonstration of their non-humanity. Does he really hope to trouble the consciences of the cops, the fascists, and the Kissingers? It all depends, Lee answers. He has been in jail; like everyone else he had his head split by billyclubs. But he always behaved the same way. "When the cop raises his stick over you, you look him straight in the eye and say 'Don't do that.' Often it has no effect. Sometimes he hesitates, then hits you. Sometimes he doesn't hit you. In any case, he keeps thinking about it afterwards."

This Country Belongs To Us

A leaflet I've brought back tells how 10,000 volunteers cleaned up the banks and the bottom of the Blackstone River (Rhode Island), returning it to swimming and boating. In a day and a half they removed 10,000 tons of detritus. They were loaned a few steamshovels and trucks gratis.

In other words, when these people wanted to recover their river, they didn't demonstrate, protest, demand, shout "what's the government doing?" They picked up shovels and wheelbarrows, and demanded steamshovels and trucks from the major polluters. In short, typical Americans start from the premise that the country belongs to them, that it will be what they make it, that it is up to them and not to the authorities to change life. The American revolution is not over.

10 May 1976